グリーンブックス

蝶類生物学英和辞典

岩野秀俊　監修
鍛治勝三　著

ニューサイエンス社

はじめに

　本書「蝶類生物学英和辞典」は、従来の単なる生物系の英和辞典とはひと味内容の異なった辞典と考えています。蝶類学、鱗翅学、昆虫学、応用昆虫学の各学問分野を主体として、それらの分野に関連する生物学や分子生物学なども包含した幅広い学問分野で、主に使用される頻度が高いと思われる英単語や英用語などを集録してあります。当初の集録用語数は、約9,000語以上にも及びましたが、私の方で校閲して取捨選択をした結果、合計で8,803語を掲載することができました。これまで昆虫学に関する辞典としては、私自身が学生時代から愛用していた素木得一博士が書かれた「昆虫学辞典(1962)、北隆館」が古くから著名な好書として知られていましたが、現在、復刻版はあるものの高価ということもあり、これに代わる昆虫学関係の普及版の新辞典が望まれていました。さらには、蝶や蛾を主体とした昆虫学に関する英和辞典となると、皆無といって過言でないほど良書が見あたらず、その必要性を痛感していたこともありました。

　そんな折り、今回、私が代表を務める「相模の蝶を語る会」の会員でもあった鍛治勝三氏より本書の出版企画の話を伺い、是非私に監修をお願いしたいという依頼がありましたので、快諾した次第です。当初は、大学の夏休み期間を利用して約2ヶ月もあれば校閲は終了するだろうと軽く考えていたのですが、いざ実際に校閲作業を行ってみると、予想以上に作業が難航して進みませんでした。一字一句ずつ、根気よく徹底的に点検するというのは、精神的にも疲れて負担が大きく、その上で莫大な時間を要する作業となってしまい、二次校正が終了したのは、冬休み明けの平成27年の1月初めの頃でした。特に、今回の校閲では、凡例にも書いてあるように、できるだけ単なるカタカナ表記は避ける方針で臨んではいたのですが、それでも昨今の分子生物学関連の語句や化学物質名などでは、英語の読みをそのままカタカナ読みして日本語に当てはめている場合が多いために、適当な和訳語が

はじめに

見あたらず、結構苦労しました。また、本書では、私の発案で日本産蝶類全275種の英名についても掲載しましたが、これも他書には見られない画期的な特徴を持った辞典になったものと自負しています。

　掲載すべき用語や語句はまだたくさんあるかと自覚していますが、ページ数の制約などもあり、今回はご了承いただければ幸いです。現代はインターネットで容易に何でも検索できる時代ではありますが、本書のような小型サイズの辞典であっても、読者の皆様方にとって存在感のある座右の書となって活用されることを祈っております。

平成27年2月

　　　　　　　　　　　　　　64(ムシ)歳を記念して　　　岩野 秀俊

目　次

はじめに（岩野秀俊）……………………………………………………………… 1 〜 2
目次 ………………………………………………………………………………… 3
凡例 ………………………………………………………………………………… 4 〜 7

蝶類生物学英和辞典

　■本編 ……………………………………………………………………………… 8 〜 163
　■付録：日本産蝶類名称の英和／和英編
　　　英和編 ……………………………………………………………………… 164 〜 169
　　　和英編 ……………………………………………………………………… 169 〜 175

おわりに（鍛治勝三）……………………………………………………………… 176 〜 178
監修者・著者略歴 ………………………………………………………………… 179

〔表紙写真〕
上段左：ヒオドシチョウ．上段中：コムラサキ．上段右：オオムラサキ♂．
中段左：キベリタテハ．中央：オオゴマダラ．中段右：オナシアゲハ．
下段左：ゴマダラチョウ．下段中：ミヤマカラスアゲハ．下段右：アカボシゴマダラ．

凡　例

1. 本編：蝶類生物学英和辞典
　●編纂方針
- 本英和辞典は、「蝶類の生物学」の学問範囲である「分類、生理、生態、形態、遺伝、発生、病理、行動、保全、採集」の各領域をできる限り網羅し、かつ、現代の研究潮流である最新の分子生物学の専門英用語もかなり網羅している。また、関連する昆虫学分野での使用に供するために、(基礎)昆虫学及び応用昆虫学の専門英用語についても補追した。収載見出し語（英用語）数は、8,800 語に及ぶ。
- 従来の専門分野の英和辞典と異なり、「専門英用語の充実」は勿論のこと、「生物学や医学などの分野での専門語の造語で基盤となる（古典）ラテン語や（古典）ギリシャ語の接頭辞」、「専門語以外の適切な英単語の選定」、「実際によく使われる英単語やフレーズ」にも力点を置いている。

　●特記事項
- 見出し語（英用語）が実際に使用される時に複数形で多用される場合でも、基本は「単数形表記」で統一した。ただし、複数形に関しては、（　）内に示し、ハイフンを用いて見出し語と共通の部分を省略したものもある。
    〔例〕antenna：(複．-nae)、触角〔ラテン語〕
    　　　census datum：(複．-ta)、調査データ

- 蝶の科や亜科の表記は、「複数形表記」とした。
    〔例〕Parnassians and Swallowtails：アゲハチョウ科
    　　　Longwings：ドクチョウ亜科

- 重要な用語で「英語」と「ラテン語」の表記がある場合には、両方を併記した。
    〔例〕palp：鬚（ひげ、しゅ）〔英語〕
    　　　palpus：(複．-pi)、鬚（ひげ、しゅ）〔ラテン語〕

- アルファベットだけの省略文字は、その和訳とフルスペルの両方を併記した。
    〔例〕A：臀脈（でんみゃく）、肛脈（Anal vein）
    　　　RFLP：制限酵素断片長多型、制限断片長多型
    　　　　　　（Restriction Fragment Length Polymorphism）

・冗長な重複の英用語は、どちらかを削除した。
　　［例］bursa copulatrix
　　　　　bursa copulatrix cell → 削除
・見出し語の綴りは、英語式より米語式を優先した。
　　［例］behaviour → behavior
　　　　　colour → color
・蝶の種名表記を統一した。
　　［例］Apollo → Apollo butterfly（「butterfly」を加筆した）
・蛾の種名表記を統一した。
　　［例］Sloane's urania → Sloane's urania moth（「moth」を加筆した）
・できるだけ単なるカタカナ表記は回避した。そのままで日本語として通用する場合や他に適語がない場合は、カタカナ表記とした。
　　［例］black list → ブラックリスト
・漢字「嚢」は、すべて「のう」とひらがな表記とした。
　　［例］bladder：香のう
・学名や蝶類学分野で必修の遺伝子名／遺伝子略名（記号）などはイタリック体で表記した。
　　［例］cactophilic *Drosophila*：好サボテン性ショウジョウバエ、
　　　　　　　　　　　サボテン好性ショウジョウバエ
　　　　　Hox gene：ホメオティック遺伝子、相同異質形成遺伝子
　　　　　in vivo：インビボ、生体内で〔ラテン語〕
・一部の和名（アレチコビトシジミ）には差別用語が残っているため、あえてそのまま掲載すると共に、今回改訂した新和名も並記掲載した。
　　○ western pygmy blue butterfly：
　　　　　　　アレチコビトシジミ（アレチコシジミ：改訂新名称）

2. 付録：日本産蝶類名称の英和／和英編
●編纂方針

・日本産の蝶類に関しては、最新の日本産蝶類標準図鑑（学習研究社発行）（白水，2006）の和名に準拠した。その図鑑に記載された迷蝶類は削除して、275種の

凡例

蝶類を選定し、その名称（和名）の英名（common English name）を記載した。

・日本産蝶類名称の英名に関しては、場合によっては亜種水準まで勘案して、英名を選定・採用した。

・「The」の表記は不要と考え、統一して削除した。

・1つの和名に対して複数の英名表記がある場合には、そのまま列記した。
　　［例］カラスアゲハ : Bianor Peacock
　　　　　カラスアゲハ : Chinese Peacock

・「和名に対して該当する英名がない種」について
　　「和名に対して該当する英名がない種」に関しては、以下の2種があり、その英名を著者が創作した。このことを明示するために、英名の後に「***」を付加した。
　　○ソテツシジミ : Cycad Butterfly ***
　　○ヤエヤマカラスアゲハ : Yayeyama Peacock ***

●特記事項

・「学名に由来した英名」について
　　蝶の名称には、学名（属名 種名）に由来した英名もある。
　　［例］ギフチョウ : Japanese Luehdorfia（属名を流用）
　　　　　ヒメウラナミジャノメ : Argus Rings（種名を流用）

・「ミスジチョウ類を表す英名」について
　　ミスジチョウ類を表す英名には「Sailer」と「Sailor」、さらには「Glider」があるが、本付録では「Sailer」に統一して使用した。
　　［例］コミスジ : Common Sailer
　　「Glider」に関しては、広く普及していると想定される場合には、それも採用した。「フタスジチョウ」がその例である。
　　○フタスジチョウ : Hungarian Glider

・「オオイチモンジ」について
　　「オオイチモンジ」に関しては、「Popular Admiral」という英名が過去の記載誤りでそのまま通用してしまったと考えられるので、ここでは「Poplar Admiral」のみを採用した。
　　○オオイチモンジ : Poplar Admiral

・「ムラサキオナガウラナミシジミ」について

凡例

　「ムラサキオナガウラナミシジミ」は、以下のような英名が付されている。
　Catochrysops strabo strabo（インド方面産）→ Oriental forget-me-not
　C. starabo luzoneris（フィリピン方面産）→ Forget-me-not
　　日本に飛来する個体群は、フィリピン方面産と考えられているので、ここでは「Forget-me-not」を採用した。
　　〇ムラサキオナガウラナミシジミ：Forget-me-not

・「ヤマトスジグロシロチョウ」について
　　「ヤマトスジグロシロチョウ」は該当する英名がなく、「エゾスジグロシロチョウ」が北海道（道東）産の別種ではあるが、（両種は、分子系統的な差異があるが、半種状態にあると考えるので、）ここでは「エゾスジグロシロチョウ」と同名の英名を採用した。
　　〇ヤマトスジグロシロチョウ：Green-veined White
　　〇エゾスジグロシロチョウ：Green-veined White

・「リュウキュウムラサキ」について
　　「Jacintha Eggfly」は、大陸亜種（学名：*Hypolimnas bolina jacintha*）の英名であるが、日本には大陸亜種も飛来するので、削除せずにそのまま残した。
　　〇リュウキュウムラサキ：Common Eggfly
　　〇リュウキュウムラサキ：Great Eggfly
　　〇リュウキュウムラサキ：Jacintha Eggfly
　　〇リュウキュウムラサキ：Varied Eggfly

以上

蝶類生物学英和辞典

－(記号)

-logy：学、論〔ギリシャ語〕
-phagy：食性、食〔ギリシャ語〕
-pter：翅〔ギリシャ語〕

0－9(数字)

11-cis-retinal：11- シス - レチナール、11- シス型レチナール
20-HE：20- ヒドロキシエクジソン(20-HydroxyEcdysone)
20-hydroxyecdysone：20- ヒドロキシエクジソン、20- ハイドロキシエクダイソン
3-hydroxykynurenine：3- ヒドロキシキヌレニン
7TMP：7 回膜貫通型タンパク質(seven TransMembrance Protein)
7TMR：7 回膜貫通型受容体(seven TransMembrance Receptor)

a

A：臀脈(でんみゃく)、肛脈(Anal vein)
a prior selection of branch：分岐の先験的選択
a prior specification of lineage：系統の先験的明確化
a tenth：10 分の 1、10 回につき 1 回
a-：無 -、非 -〔ギリシャ語〕
AAT：アスパラギン酸アミノ基転移酵素、アスパラギン酸アミノトランスフェラーゼ(Aspartate AminoTransferase)
ab-：離れて -、外側の -、反対側の -〔ラテン語〕
ab.：異常型(aberrant form)
abandonment：放棄、中止
abaxial：反軸側(はんじくそく)の、背軸側の
abbreviation：略記
abdomen：腹部〔ラテン語〕
abdomen mass：腹部重量
abdominal defensive gland：腹部防御腺、腹性防御腺
abdominal ganglion：(複 .-glia)、腹部神経節〔ラテン語〕
abdominal proleg：腹脚(ふくきゃく)、腹部の脚
aberrant：異常型、異常
aberrant form：異常型
aberration：変体、変性種、異常
abide：守る、遵守する、とどまる、待つ
ability of to-and-fro movement：往復移動の能力
abiotic environment：非生物的環境
abiotic factor：非生物的要因
ablative operation：除去手術
abnormal activity：異常行動
abnormal climate：異常気候
abnormal wing：異常な翅
abolish：抑制する、機能しなくなる
abort：流産する
abrupt change：突然の変化
absolute tautonymy：完全同語反復
absorb：吸収する
absorbance spectrum：吸収スペクトル
absorbance spectrum maximum：吸収スペクトルの極大波長、最大吸収スペクトル

absorbent paper：濾紙
absorbent tissue：吸収性のティッシュ
absorption：吸収
absorption efficiency：吸収効率
absorption maximum：吸収極大
absorption peak：吸収ピーク
absorption spectrum：吸収スペクトル
absorption value：吸収値
absorption wavelength region：吸収波長域
abstract：要旨、概要、摘要
abundance：多量、多数、豊富、存在量、個体数、個体数量
abundance of a species：種の個体数
abuse of species diversity index：種多様性指標の誤用、種多様性指数の誤用
accept：受理、アクセプト
acceptability：受容性
acceptable threshold：容認可能な限度
accession number：受入番号
accessory genital gland：生殖器付属腺
accessory gland：付属腺
accessory pigment cell：補助色素細胞
acclimation：順応、順化、馴化（じゅんか）
acclimation period：馴化期間
acclimatization：順化、馴化（じゅんか）
acclimatization process：順化過程
accrue：生じる、手に入れる
accumulated temperature：積算温度
accumulated value：積算値
accumulation：蓄積
accumulation curve：累積曲線
accumulation of neutral mutation：中立的突然変異の累積
achene dispersal distance：痩果の分散距離

achromatic vision：明暗視、全色盲
acknowledgements：謝辞
ACO：アリコロニー最適化法（Ant Colony Optimization）
acoustic stimulus：音刺激、音響刺激
acquired ammonia：獲得されたアンモニア
action effect：作用効果
action spectrum：作用スペクトル
activate：活性化する、活動的にする
activation stimulus：活性化刺激
active brain：活性脳
active phase：活動相、活動位相
active season：活動時期、活動季節
active stage：活動期
activity locus：活動点、活動場所
activity radius：行動半径、行動範囲
activity range：行動範囲
actual mating probability：実際の交尾確率、実効的交尾確率
acute：急性の
adaptation：適応、順応
adaptative learning：順応学習、適応学習
adaptive convergence：適応的収斂
adaptive differentiation：適応的分化
adaptive dispersal：適応的分散
adaptive evolution：適応進化
adaptive introgression：適応的遺伝子移入
adaptive management：順応的管理
adaptive mutation：適応的突然変異
adaptive novelty：適応的新奇性、適応的新規性
adaptive radiation：適応放散
adaptive response：適応的反応
adaptive seasonal polymorphism：適応的季節多型

adaptive significance：適応的意義、適応上の意義

adaptive system：適応様式

adaptive trait：適応形質

additive：加法的な、付加的な、相加的な

additive genetic variance：相加的遺伝分散

additive genotypic value：相加的遺伝子型値

additively：加法的に、相加的に

adenosine triphosphate：アデノシン三リン酸

adhesion of the barbed and sticky bristle：棘のある粘着性剛毛の粘着力

adhesive tape：粘着テープ

adipokinetic hormone：脂質動員ホルモン

adjacent distribution：隣接分布地域

adjacent lane：隣接レーン

adjoin：接合する

admiral butterfly：アカタテハ属の蝶

Admirals and Relatives：イチモンジチョウ亜科

admixture：混合

adopt：採用する

adoption：採用

adult：成虫、成体

adult diapause：成虫休眠

adult eclosion：羽化した成虫、成虫の羽化

adult emergence season：成虫出現期、成虫羽化時期

adult form：成虫形状

adult longevity：成虫の寿命

adult mass：成虫の体重

adult maturation：成虫の成熟

adult morphology：成虫の形態

adult overwintering：成虫での越冬、成虫越冬

adult size：成虫のサイズ

adult stage：成虫期

adult wing：成虫翅

adult-female：成虫雌

adulthood：成虫期

adventitious bud：不定芽

adversity：逆境

adversity selection：逆境淘汰

adzuki bean borer moth：アズキノメイガ

aedeagus (ae)：交尾棍、エデアグス、挿入器

aedoeagus (aedeagus) (ae)：エデアグス、挿入器

aequorin：エクオリン、イクオリン（クラゲの発光タンパク質）

aequorin gene：エクオリン遺伝子、イクオリン遺伝子

aerial hawking bat：ヒメホリカワコウモリ

aerodynamics：空気力学

aeropyle：気孔

aestivate：夏眠する〔米語〕

aestivation：夏眠、夏休眠〔米語〕

aff.：の近似（affinis）、類似、近似種

affected native species：影響を受ける在来種

AFLP：増幅断片長多型（Amplified Fragment Length Polymorphism）

African migrant butterfly：ミズアオシロチョウ

Afrotropical region：熱帯アフリカ区、エチオピア区

agarose gel：アガロースゲル

age of adult：成虫の年齢

age specific life table：齢別生命表、年齢別生命表

age-related sexual receptivity：日齢に関

係する交尾受容性

age-related site fidelity：日齢に関係する出生場所固執性、日齢に関係する生息場所執着性

Agency for Cultural Affairs：文化庁(日本)

aggregate：集群、集合(体)

aggregation：集団、集まり

aggregation pheromone：集合フェロモン

aggregative feeding：集団摂食

AGH：造雄腺ホルモン(Androgenic Gland Hormone)

agility：機敏、機敏性

agricultural landscape：農業景観

agriculture：農耕、農業

AIC：赤池情報量基準(Akaike's Information Criterion)

air hole：通気孔

air temperature：気温

Akaike's information criterion：赤池情報量基準

aktionsraum：行動圏

alarm pheromone：警報フェロモン

alata：有翅型

albinic：白化の

albino form：白化型

alcoholic Bouin's fixative：アルコールブアン固定液

alien species：外来種、移入種、帰化種

alighting：着陸

align：並べる、整列させる

alignment：整列、アラインメント

alignment with sequence：配列のアラインメント

alkaloid：アルカロイド

all year：年中、一年中

all-female brood：すべて雌の同腹仔

all-female-producing matriline：すべて雌を産生する母系群

all-male cage：すべて雄が入っているケージ

allantoic acid：アラントイン酸

allantoin：アラントイン

allatostatin：アラトスタチン

allatotropin：アラトトロピン

Allee effect：アリー効果

Allee's effect：アリー効果

allele：対立遺伝子、アレル

allele difference：対立遺伝子の違い

allelic size variation：対立遺伝子のサイズ変異

allelic variant：対立遺伝子変異体

allelic variation：対立遺伝子変異

allelo-：相互の -、お互いの -〔ギリシャ語〕

allelochemical：アレロケミカル、他感作用物質、他感物質

allelopathy related compound：アレロパシー関連化合物

allied species：近縁種

allo-：異 -、別 -〔ギリシャ語〕

allochronic：異時性、異時的な

allochronic isolation：時間的隔離、異時的隔離

allochthonous element：他生的な要素

allochthonous species：他生的な種、外来種

allomelanin：アロメラニン

allometry：アロメトリー、相対成長

allomimesis：隠蔽的異物擬態

allomone：アロモン

allopatric：異所性の、異域性の

allopatric and genetic speciation：異所的な遺伝的種分化

allopatric speciation：異所的種分化
allopatry：異所性
allotype：アロタイプ、別模式標本
allozyme：アロザイム
allozyme allele frequency：アロザイム対立遺伝子頻度
allozyme diversity：アロザイム多様性
allozyme electrophoresis：アロザイム電気泳動
allozymic differentiation：アロザイム分化
ally：近縁種
alongside：並んで、一緒に
alorium：爪間盤(板)、アロリウム
alpine：高山性、高山性の
alpine butterfly：高山蝶
alpine meadow：高山性草原
alpine organism：高山性生物
alpine tundra：高山ツンドラ
alpine tundra grassland：高山ツンドラ草地
alpine zone：高山帯
alternate phenotype：交互に発現する表現型
alternating season：交互に入れ替わる季節
alternating seasonal environment：交互に入れ替わる季節環境
alternative adult phenotype：交互に発現する成虫の表現型
alternative allele：代替の対立遺伝子
alternative model：代替モデル
alternative temperature environment：交互に入れ替わる温度環境
altitude：高度、標高
altitudinal optimum：高度の最適
altitudinal shift：高度シフト、高度変化

altruistic behavior：利他(的)行動
am-：周囲 -〔ラテン語〕
amateur naturalist：アマチュアの自然科学研究者、アマチュア研究者
Amazon basin：アマゾン川流域
Amazonian butterfly：アマゾンの蝶
amb-：周囲 -〔ラテン語〕
ambient condition：周囲条件、環境条件
ambient temperature：周囲温度、外気温
ambiguity：多義性、不確かさ
ambiguous electromorph：多義的電気泳動パターン
American copper butterfly：アメリカベニシジミ
American snout butterfly：アメリカテングチョウ
ametaboly：無変態
amibient temperature：周囲温度
amino acid：アミノ酸
amino acid change：アミノ酸置換、アミノ酸変異
amino acid difference：アミノ酸配列の違い
amino acid replacement：アミノ酸置換
amino acid residue：アミノ酸残基
amino acid secretion：アミノ酸の分泌物
amino acid site：アミノ酸部位
amino acid substitution：アミノ酸置換
amino acid substitution frequency：アミノ酸置換頻度
amino acid substitution model：アミノ酸置換モデル
amino-acid sequence：アミノ酸配列
ammonia：アンモニア
ammonia ingestion：アンモニア摂取
ammonia uptake：アンモニアの吸収、ア

ンモニアの摂取
ammonium：アンモニウム
ammonium chloride：アンモニウム塩化物、塩化アンモニウム
amnion：羊膜
among-taxa component：分類群間の成分
amount of assimilation：同化量
amount of blue bilin：青色ビリン量
amount of light：光量
amount of light energy：光エネルギー量
amount of seminal protein：精液中のタンパク質の量
AMOVA：分子分散分析（Analysis of MOlecular VAriance）
amphibian：両生類
ample anatomical datum：（複．-ta）、十分な解剖データ
ample artificial nectar：たっぷりな人工果汁
ample opportunity：十分な機会
amplicon length variation：アンプリコン長の変異
amplified DNA fragment：増幅DNA断片
amplified fragment length polymorphism marker：増幅断片長多型マーカー
Amur type：アムール型
anabolism：同化
anagram：アナグラム
anal：肛門（こうもん）、臀部、臀脈（でんみゃく）、A脈、肛側
anal angle：肛角部
anal fold：肛門褶
anal proleg：肛門部の脚、尾脚
anal vein：臀脈（でんみゃく）、A脈
analysis of covariance：共分散分析

analysis of molecular variance：分子分散分析
analysis of stage-structured population：ステージ別個体群解析
analysis of variance：分散分析
anatomically：解剖学的に
anatomy：体の構造、解剖
ancestral bat detector：先祖伝来のコウモリ探知器
ancestral heliconiine group：祖先のドクチョウグループ
ancestral larval foodplant：祖先的な幼虫の寄主植物
ancestral node：祖先ノード
ancestral state reconstruction：祖先状態再構成、祖先状態復元、先祖状態再構築
anchor：固定する、アンカー
anchor locus：アンカー遺伝子座、固定遺伝子座
ANCOVA：共分散分析（ANalysis of COVAriance）
andesitic：安山岩質の
androconial organ：発香器官
androconial scale：発香鱗
androconial secretion：発香器官分泌物
androconial system：発香器官系
androconium：香鱗、発香鱗
androgenic gland：造雄腺
androgenic gland hormone：造雄腺ホルモン
anepisternum：上前腹板
angiosperm：被子植物
angiosperm flower：被子植物の花
angular：とがった、角張った、鋭角的な
angular spot：角張った斑点

anhydrobiosis：乾燥休眠、アンヒドロビオシス、乾眠
animal：動物
animal droppings：獣糞
animal speciation：動物の種分化
anion：アニオン、陰イオン
anisotropy：異方性、不均等性
annelifer：アネリフェル
annellus：薄膜の鞘
annonaceous acetogenin：アノナセウスアセトゲニン
annotation：注釈、注解
annual change：周年変化
annual cycle of cool-warm climate：寒暖気候の周年サイクル
annual cycle of wet-dry climate：乾湿気候の周年サイクル
annual herbaceous plant：一年生草本
annulated：環化した、環のある
anonymous：匿名の、名のない
anonymous reviewer：匿名の査読者
anonymous work：匿名の著作物
ANOVA：分散分析（ANalysis Of VAriance）
ant：アリ、蟻
ant colony optimization：アリコロニー最適化法
ant-enriched habitat：アリが豊富な生息地
antagonistic pleiotropy：拮抗的多面発現
Antarctic region：南極区
ante-：前の-、前に-〔ラテン語〕
antenna：（複．-nae）触角〔ラテン語〕
antenna shape：触角の形状
antennapedia：アンテナペディア遺伝子
anterior：前、前部の、前側の、前方の〔ラテン語〕
anterior branch：前分枝
anterior cross vein：前横脈
anterior cubitus vein：前肘脈、Cu脈、中脈（ちゅうみゃく）
anterior edge：前縁、前端
anterior eyespot：前方の眼状紋
anterior location：前方の位置
anterior margin：前縁（部）
anterior tentacle organ：前部伸縮突起、前方伸縮突起
anterior-posterior axis：前後軸、AP軸、前縁-後縁軸
antero-：前部-〔ラテン語〕
anteroposterior axis：前後軸、AP軸、前縁-後縁軸
anteroposterior scale ring：前縁-後縁方向に取り巻く鱗粉リング
anteroposteriorly：前縁-後縁方向へ、前後方向に
anthophyta：被子植物門
anthoxanthin：アントキサンチン
anthropogenic climate warming：人為的な気候温暖化
anthropogenic habitat：人為的に改変された生息地
anthropogenic habitat change：人為的な生息地変化
anti-：反-、不-〔ギリシャ語〕
anti-freezing substance：凍結防止物質
antiaphrodisiac：抗催淫剤
antibiosis：抗生作用
antibiotic：抗生物質
antibiotic treatment：抗生物質処理
antibiotic-containing diet：抗生物質入りの飼料
antibody：抗体

antifeedant：摂食阻害物質
antifreeze protein：不凍タンパク質
antigen：抗原
antioxidant enzyme：抗酸化酵素
Antp：アンテナペディア遺伝子（*Antennapedia* gene）
anucleate spermatozoon：無核精子
any time diapause：随時的休眠
AP axis：前後軸、AP軸、前縁 - 後縁軸（Anterior-Posterior axis）
ape：類人猿
apex：頂部、翅頂部、(翅)端部、翅端、稜、先端(方)の
aphid：アブラムシ
aphrodisiac：催淫剤
aphrodisiac pheromone：催淫性フェロモン
apical：頂室、翅頂室、(翅)端室、頂端の、翅頂部帯、翅端部、先端(方)の〔ラテン語〕
apical cell：頂室、端室、末端細胞
apical side：頂端側
apical spot of cell：中室端斑
apiculture：ミツバチ飼育、養蜂業
apiculus：ふ先、先端の細い毛
apo-：離れて -、別れて -〔ギリシャ語〕
Apollo butterfly：アポロウスバシロチョウ
apomixis：アポミクシス、無融合生殖
apomorph：後生的進化形質、後天的新形質
apoptosis：アポトーシス、プログラム細胞死
aposematic：警告する、警戒色の
aposematic caterpillar：警戒色の幼虫
aposematic coloration：警告色、警戒色
aposematic insect：警告性昆虫
aposematism：警告誘示、警告戦略

apparency：顕示性(度)
apparent competition：見かけの競争
apparent decline：明白な衰亡
apparent fusion of eyespot：眼状紋の外見上の融合
apparently dead pupa：外見的には死んだ蛹
appearance：外見、発現、外観
appendage：付属肢(ふぞくし)、付属器、付属突起
apple：リンゴ
applied entomology：応用昆虫学
appreciate：正しく認識する、〜の真価（性質、差異）を認める
approach：接近
approach probability：接近確率
approximate：近づける、近似の
approximately：おおよそ
aptera：無翅型
apterous：無翅の
apyrene sperm：無核精子
apyrene spermatozoon：無核精子
AQS：単為生殖による女王位継承システム（Asexual Queen Succession）
aquatic invertebrate assemblage：水生無脊椎動物群集
arbitrary：任意の
arbitrary combination of letter：文字の任意組合せ
arbitrary threshold：任意の閾(いき)値、任意閾
arbor：あずまや、日よけの場所
Archaeognatha：イシノミ目、古顎目
archival preservation：アーカイブ保存、公文書史料保管
arctic：北極(圏)の

arctic species：北極圏種

arctic zone：北寒帯、北極帯

Arcto-Tertiary element：第三紀北極要素、第三紀周北極要素、第三紀周極要素

area of overlap：重複地域、重複する区域

area-based accumulation curve：面積に基づく累積曲線

area-wide integrated pest management：広域的総合的害虫管理、広域総合的害虫管理

areal extent：空間範囲、範囲

areola：（複．-lae）、小室〔ラテン語〕

areole：小室〔英語〕

arid grassland：乾燥地帯

arid region：乾燥地

arid tropical：乾燥熱帯性

arid zone：乾燥地帯

Aristolochiaceae：ウマノスズクサ科

arithmetic progression：等差数列

arm race：軍拡競争

army：群れ、大群

arolium：爪間盤（そうかんばん）、アロリウム

arrangement：配列、並び方

arrangement pattern：配置パターン

arrested development：発育停止、発育遅滞、発育遅延

arrested embryogenesis：阻害された胚形成、胚発達停止

arrival of a new species：新たな種の出現

arrow-shaped marking：矢印状の斑紋

arthropod：節足動物（せっそくどうぶつ）

arthropoda：節足動物門

article：条項、論説、品目

artificial barrier：人工障壁

artificial diet：人工飼料

artificial environment：人工的な環境

artificial food：人工飼料

artificial nectar：人工果汁

artificial selection：人為選択

artificial short-day：人工短日

artificial substrate：人工基質、人工的な培養基

artificial transfer：人工的な転換

artificially induced species：人為導入種

arylphorin：アリルフォリン、アリルホリン遺伝子

aseasonal migration：非季節的移住

aseasonal summer frost：季節外れの夏に降りた霜

asexual queen succession：単為生殖による女王位継承システム

asparagine-to-serine substitution：アスパラギンからセリンへの置換

aspartate aminotransferase：アスパラギン酸アミノ基転移酵素、アスパラギン酸アミノトランスフェラーゼ

aspect ratio：アスペクト比、縦横比

assemblage：群集、集団

assemblage-level thinning hypothesis：群集水準の間伐説、群集レベルの間伐説、群集レベルの間引き説

assessment protocol：評価プロトコール、評価計画案

assimilate：同化する

assimilation：同化、同化作用

association between color and preference：色と選好性の関連

association study：関連解析、関連研究

associative learning：連合学習

assortative mate preference：同類交配選好

assortative mating：同類交配、選択交配
asymmetric：非対称の
asymmetric gene flow：非対称遺伝子流動
asymptote：漸近線
asymptotic estimator：漸近的推定量
asymptotic richness：漸近的種数、漸近種数
asymptotic richness estimator：漸近種数推定量
asymptotically：漸近的に
asymptotically equivalent：漸近的に等価
asynchronous emergence：不斉一発生、発生の不斉一性
atlas moth：ヨナクニサン
atmospheric phenomenon：気象現象
atomic coordinate：原子座標
ATP：アデノシン三リン酸（Adenosine TriPhosphate）
attach distally：末端側で結合する、末梢側で結合する
attack：攻撃（する）
attacker：捕食者
attain：達成する
attenuate：減衰させる、減衰する、弱める
attenuated feminization：減衰された雌化
attenuated feminizing activity：減衰された雌化行動
attenuation：減衰、抑制
attract：誘引する
attractant：誘引物質
attributable：帰すことができる、基因する
atypical：異常な、定型的でない
auctorum：著者たちの〔ラテン語〕
audible sound：聞き取れる音
audiogram：オージオグラム、聴力図
audiogram experiment：聴力図実験、オージオグラム実験
audiogram procedure：聴力検査手順、聴力図作成手順
auditory：聴覚（の）
auditory nerve：聴覚神経
auditory response：聴覚応答、聴覚反応、聴性反応
auditory sense：聴覚
auditory threshold：聴覚閾（いき）値、聴覚閾、最小可聴値
augment：増大（する）
augmentation：放飼増強法
augmented ultraviolet color vision：増強された紫外線光に対する色覚
auricular：耳の、聴覚の
Australian region：オーストラリア区
autapomorphy：固有派生形質
author：命名者、著者
authorisation procedure：承認手続き
auto-：自己 -、同種 -、自動 -〔ギリシャ語〕
autoapomorphy：固有派生形質
autoclaved：高圧滅菌された
automated purification instrument：自動精製装置
automimicry：種内擬態、自己擬態
automixis：オートミクシス、自家生殖
autonomous oscillator：自律振動体
autotrophy：独立栄養、自家栄養
autumn form：秋型
autumn morph adult：秋型成虫
autumn rain：秋雨
availability：適格性、利用価値
availability of habitat：生息地の利用可能性
available name：適格名、適用名

available nomenclatural act：適格な命名法的行為
available work：適格な著作物
average：平均、平均値
average density：平均密度
average sound pressure level：平均音圧レベル
avian predator：鳥捕食者
avoidance：忌避作用、忌避性
avoidance behavior：忌避行動
AW-IPM：広域的総合的害虫管理、広域総合的害虫管理（Area-Wide Integrated Pest Management）
await：待ち構える
axial：軸側（じくそく）
axiality：軸性
axillary sclerite：腋節片
axis：軸、中心線〔ラテン語〕
axon：軸索
axon terminal：軸索末端

b

BAC：細菌人工染色体、バクテリア人工染色体（Bacterial Artificial Chromosome）
BAC clone identification：BACクローン同定
back：背、背面
back and side：背部と体側
back out of eye：目の後ろあたりで
backcross：戻し交配、戻し交雑
backcross brood：戻し交配の同胞
backcross family：戻し交配の家系
backcross male preference：戻し交雑の結果の雄の選好性
background activity：背景活動
background extinction：背景絶滅、バックグラウンド的絶滅
background noise：背景雑音、暗騒音
backward trajectory analysis：後退流跡線解析
bacterial endosymbiont Wolbachia：細胞内共生細菌ボルバキア
bacterium：（複 .-ia）、細菌、バクテリア
baculovirus：バキュロウイルス
bagworm：ミノムシ
balance of nature：自然のバランス
balancing selection：平衡選択
ball-like cluster of hair：ボールのようなかたちをした毛束
band：帯、模様、縞状バンド
band region of adult wing：成虫翅の帯領域、成虫翅の帯部分
band string：帯糸（たいし）
banded orange butterfly：オビモンドクチョウ
banding pattern：バンドパターン
bandwidth：帯域幅
banker plant system：バンカー法
barely significant reduction：まれに有意となる減少
basad：基部側へ
basal：基部帯、基部（方）の
basal band：基部の縞状バンド
basal cell：基底細胞
basal group：基底グループ、基部群
basal node：基底ノード、基部の結節
base：基角、塩基
base change：塩基置換
base of forewing：前翅基部
base of tympanal chamber：鼓膜室の基部
base pair：塩基対

base sequence：塩基配列
base substitution：塩基置換
base substitution number：塩基置換数
basement membrane：基底膜
Basic Local Alignment Search Tool：BLAST 検索
basic skeleton：基本骨格
basin：盆地、流域、入り江
basking：日光浴
bat detection：コウモリ探知
bat evasion：コウモリからの逃避
batch of egg：卵塊
Bateman's principle：ベイトマンの原理
Batesian mimicry：ベイツ型擬態
battle：闘争
Bayes posterior probability：ベイズの事後確率
Bayesian algorithm：ベイズのアルゴリズム、ベイズの問題解決手法
Bayesian phylogenetic analysis：ベイズ系統解析、ベイズ法による系統解析
bcd gene：ビコイド（*bicoid*）遺伝子
beak mark：ビークマーク、鳥の嘴型
bee：蜜蜂
beech band：ブナ帯
beginning of light period：明期開始
behavior：行動、活動
behavior level：行動レベル
behavior of female refusing male：雌の交尾拒否行動
behavioral：行動によるもの、行動の
behavioral and neurophysiological response：行動的・神経生理的応答
behavioral approach：行動学的アプローチ
behavioral characteristic：行動特性

behavioral ecology：行動生態学
behavioral incompatibility：行動的不一致
behavioral isolation：行動的隔離
behavioral protandry：行動学的プロタンドリー、行動面での雄性先熟
behavioral response：行動的応答、行動的反応
behavioral syndrome：行動シンドローム
bellows：蛇腹
below surface：裏翅、裏面
belt shaped area：帯状地域
benchmark dataset：ベンチマークデータセット、標準データセット
benchmark seedbank dataset：基準種子銀行データセット、基準シードバンクデータセット
beneath：真下に
beneficial effect：有益な効果
beneficial insect：益虫
beneficial trait：有益な形質
benefit：利益、収益
benign introduction：保全的導入
benthic invertebrate assemblage：河口底生無脊椎動物群集
benthic marine mass collection：海洋底生の大量採取品
berry：木の実、果実
best threshold：最適な閾（いき）値、最適閾値
bet-hedging：危険分散、二股賭け戦略、両掛け戦略
bet-hedging adaptation：危険分散適応、両掛け適応、両掛け戦略適応
bet-hedging effect：危険分散の効果
bet-hedging strategy：危険分散戦略、両掛け戦略

beta-alanine (β-alanine)：β-アラニン、ベータアラニン
beta-carotene (β-carotene)：β-カロチン、ベータカロチン
between-morph courtship：異色の翅間での求愛
bi-：2-、二-、双-、2倍-〔ラテン語〕
bias：バイアス、先入観、偏り
bibliographic reference：参考文献
bicuspid apex of valva：交尾弁二尖、交尾弁縁棘
biennial：二年ごとの、二年生の、越年生の
bilin：ビリン
bilin-binding protein：ビリン結合タンパク質
biliverdin IX：ビリベルジン IX
biliverdin type：ビリベルジンタイプ
billion：十億の
bimodally：二峰的に、二方式に
binocular microscope：双眼顕微鏡
binomen：（複．-mina）、二語名〔ラテン語〕
binomial datum：（複．-ta）、二項データ
binomial nomenclature：二名法、二名式命名法、二名式用語体系
binomial regression model：二項回帰モデル、二項分布回帰モデル
binominal name：二語名
binominal nomenclature：二名法
bio-：生物-〔ギリシャ語〕
bioassay：生物検定
bioateral symmetry：左右相称
biochemical evolution：生化学的進化
biochemistry：生化学
biocoenosis：群集
biodiversity：生物多様性、生物学的多様性
biodiversity assessment：生物多様性評価、生物の多様性評価
bioeconomics：生物経済学
biogeographical region：生物地理区
biogeography：生物地理学
biological characteristic：生物学的特徴
biological clock：生物時計、体内時計
biological control：生物（学）的防除
biological diversity：生物多様性
biological factor：生物学的要因、生物学的因子
biological infrared imaging：赤外光による生体イメージング
biological infrared sensing：赤外光による生体センシング
biological invasion：生物学的侵入
biological isolation：生物的隔離
biological species：生物種、生物学的種
biological species concept：生物学的種概念
biologically：生物学的に
biologically important expansion：生物学的に重要な拡大
bioluminescent jellyfish：発光クラゲ
biomass：生物（体）量、バイオマス
biome：生物群系、バイオーム
biomimetics：バイオミメティク、バイオミメティックス
biosphere：生物圏
biosynthesis：生合成
biosynthetic pathway：生合成経路
biosystematics：生物系統学、生物分類学
biota：生物相
biotic factor：生物的要因
biotic inventory：生物目録、生物資源目録
biotic survey：生物調査、生物資源調査
biotope：小生活圏、ビオトープ

biotype：バイオタイプ、生物型、遺伝因子型
biped：二足動物、二足を有する
bird：鳥
bird droppings：鳥の糞
bird vocalization：鳥のさえずり、鳥の鳴き声
birdcall：鳥の鳴き声
birdwing butterfly：トリバネアゲハ
bistability phenomenon：双安定現象、二相安定現象
bithorax：双胸遺伝子
biting：噛む（虫が）
bivoltine：二化性の、二化性
bivoltine population：二化性集団
bivoltine zone：二化性地域、二化地帯
bizarre：奇異な、奇怪な、風変わりな
black and buff eyespot：暗黄褐色の眼状紋
black annulus：黒環
black dorsal hindwing：黒色の背側後翅
black dot：黒丸、黒点
black flanking orange：橙色で囲まれた黒色
black list：ブラックリスト
black patch：暗色斑
black pigment：黒色色素
black pupation board：黒色の蛹化台紙
black ring：黒環
black scale：黒色鱗粉
black streak：黒条
black stripe on outer edge：外縁黒帯
black stripe size：黒帯幅
black swallowtail butterfly：黒色系アゲハ
black-ringed：黒く縁どられた
black-vein：黒条

blackish brown：黒味をおびた褐色
blackish upperside：黒っぽい翅表
bladder：香のう
BLAST comparison：BLAST 比較（Basic Local Alignment Search Tool）
blastoderm：胚盤葉、胞胚葉
Blattodea：ゴキブリ目、網翅目
blemish：傷、傷つける
blood：体液、血液
blue bilin accumulation：青色ビリンの蓄積
blue bilin pigment：青色ビリン色素
blue bilin production：青色ビリン色素の生成
blue jay：アオカケス
blue light：青色光
blue metalmark butterfly：シジミタテハ
blue morpho butterfly：メネラウスモルフォ
blue pigment：青色色素
blue spectral shift：青色のスペクトル領域へのシフト、青色スペクトル範囲へのシフト、青色スペクトルシフト
blue-banded female model：青色翅の雌モデル
blue-green coloration：青緑色化
blue-shift：青色側へのシフト、青方偏移、ブルーシフト
blue-shifted lineage：青色シフト系統
Blues：ヒメシジミ亜科、シジミチョウ科
blunt：鈍い、先のとがっていない
BM：大英博物館（British Museum）
BmN4：カイコ（*Bombyx mori*）の卵巣由来の樹立培養細胞株
BMNH：ロンドン自然史博物館（British Museum (Natural History)）
body：胴、体、胴体

body color：体色
body color change：体色変化
body component：体の器官
body fluid：体液
body morphology：体部の形態、体の形態
body of female：雌の体
body plan development：ボディプランの発育
body resource：体部の資源
body size：体の大きさ、体サイズ
body size indicator：体の大きさの指標、体サイズ指標
body structure：体構造
body surface area：体表面積
body temperature：体温
bog：湿原、沼地
bog butterfly：湿原性チョウ
bold：はっきりした
bold line：太線
bombyxin：ボンビキシン
bona fide scientific purpose：本物の科学的目的
bonanza：幸運、大当たり
bone：骨
Bonferroni correction：ボンフェローニ（の）補正
boom and bust cycle：好景気 - 不景気サイクル
boost：増加する
bootstrap：ブートストラップ
bootstrap probability：ブートストラップ確率
bootstrap replicate：ブートストラップ複製
bootstrap support：ブートストラップサポート

borax solution：ホウ砂溶液
border：外縁、縁どり、国境、境界
border control：国境検査
border ocellus：辺縁部の単眼
boreal species：北方系種、寒帯種
both flanking marker：両側の隣接マーカー
both sexes：雌雄両方、両性
bottleneck：ボトルネック、瓶首
bottleneck effort：ボトルネック効果、瓶首効果
boundary：境界、境界線
boundary layer：境界層
bourgeois strategy：ブルジョワ戦略
bovine rhodopsin：ウシロドプシン
bovine template：ウシテンプレート
bp：塩基対（単位）（base pair）
brachypter：短翅型
brachypterism：短翅
brachypterous form：短翅型
brain：脳〔ラテン語〕
brain, prothoracic gland, corpus allatum, corpus cardiacum complex：脳 - 前胸腺 - アラタ体 - 側心体の複合体
brainless animal：除脳動物
brainless diapausing-pupa：除脳休眠蛹
brainless pupa：除脳蛹
branch-site model of selection：自然選択の分岐点モデル
branch-site test of selection：選択の分岐点テスト
branched spine：枝分れしたトゲ
brand：性標、性斑
break mark：咬み跡
breathing pore：気門、気孔
breeding habitat：繁殖地、繁殖場所

breeding season：繁殖期
breeding value：育種価
brief light pulse：短時間の光パルス
bright wing pattern：鮮やかな翅の色彩
bright yellowish green pupa：明黄緑色の蛹
brightness：明度
brightness processing：明度処理、輝度処理
brightness-contrast vision：明暗対比型視覚
bring about：引き起こす、もたらす
brink of extinction：絶滅の瀬戸際
bristle：剛毛
British Museum (Natural History)：ロンドン自然史博物館（BMNH）
British Museum：大英博物館（BM）
broadcast：放送、ブロードキャスト
broadleaved woodland：広葉樹林地帯
broken band：不連続な帯
broken line：不連続な帯、破線
brood：同腹の仔、ブルード、同腹、同腹仔
brood comparison：同腹仔の比較、ブルード比較
brood-dependent variation：同腹仔依存の変動
brother-sister mating：兄妹交尾、同胞交配
brown butterfly：ジャノメチョウ
brown leaf litter：褐色の腐葉（層）
brown marking：褐色斑紋
brown pigment：褐色色素
brown planthopper：トビイロウンカ
brownish：褐色がかった
Browns：ジャノメチョウ科
browsing：若葉を食べる

bruchid beetle：ヨツモンマメゾウムシ
brush：雑木林、低木林、やぶ
brush-footed：刷毛足の
Brush-footed Butterflies：タテハチョウ科
BSC：生物学的種概念（Biological Species Concept）
buckeye butterfly：アメリカタテハモドキ
buddleia：フジウツギ、ブッドレア
buffer strip：緩衝帯
Bulletin of Zoological Nomenclature：動物命名法紀要
bump：こぶ、瘤
burnin：バーンイン（ベイズ法関連用語）
bursa copulatric：交尾のう
bursa copulatrix：交尾のう
bursicon：ブルシコン
butter-colored fly：バター色のハエ・アブ
butterflies：蝶類
butterfly：蝶
butterfly bush：蝶の木（フジウツギ科ブッドレアの別称）
butterfly distribution：蝶の分布
butterfly fauna：蝶相、チョウ相
butterfly garden：バタフライガーデン、蝶園
butterfly genome：蝶のゲノム
butterfly morphology：蝶の形態（学）
butterfly pigment：蝶の色素
butterfly population：蝶の個体群
butterfly transect：蝶の観察路、蝶の観察地
butterfly-co-rotating：蝶の共回転
bx：双胸遺伝子（*bithorax* gene）
by-product：副産物
by-product of ecological selection：生態（学的）選択の副産物

by-product of ecological shift：生態転換の副産物
by-product of sexual selection：性選択の副産物
bya：十億年前（billion years ago）
bylaw：条例、内規

c

C：前縁脈（Coastal vein）
cabbage：キャベツ
cabbage white butterfly：モンシロチョウ
cactophilic *Drosophila*：好サボテン性ショウジョウバエ、サボテン好性ショウジョウバエ
caffeine-sensitive neuron：カフェイン感受性ニューロン、カフェイン感受性神経細胞
cage：ケージ、籠（かご）
cage complex：籠複合体、飼育ケージ複合体
Cairns birdwing butterfly：メガネトリバネアゲハ（旧称）、プリアムストリバネアゲハ（ケアンズはオーストラリア北東部の地名で、生息地の一つ）
calcareous grassland：石灰土壌の草地
calcareous soil：石灰質土壌
calcium concentration：カルシウム濃度
calcium imaging assay：カルシウムイメージング（検定）法
calcium-dependent luminescent protein：カルシウム依存的発光タンパク質
calculated migration：目標移動
calibrate：調節する、調整する
California dog-face butterfly：カルフォルニアイヌモンキチョウ
calling sound of bird：鳥の呼ぶ音

Camberwell beauty butterfly：キベリタテハ
camouflage：擬態、隠蔽、カムフラージュ
camouflaged ventral side of wing：擬態した翅の腹側
campaniform sensillum：鐘状感覚子
candidate gene：候補遺伝子
candidate gustatory receptor gene：味覚受容体候補遺伝子、味覚受容体遺伝子候補
candidate spectral tuning site：スペクトル調整部位候補
cannibalism：共食い（ともぐい）、カニバリズム、種内捕食
canopy：林冠、樹冠、天蓋
cap：帽鞘（ぼうしょう）、キャップ
capillary：キャピラリー、毛細管、毛管
captivating hobby：魅力たっぷりな趣味
captive breeding：飼育環境下での繁殖、人工繁殖
captive condition：飼育条件、捕獲下条件
captive-bred rare species：飼育環境で繁殖された希少種
captive-bred stock：飼育環境で繁殖されたストック
captivity, in：飼育されている
carbohydrate：炭水化物、糖質
carbohydrate metabolism：炭水化物代謝
carboxylesterase：カルボキシルエステラーゼ
cardenolide：カルデノリド、強心配糖体
cardenolide concentration：カルデノリド濃縮
cardenolide content：カルデノリド含有
cardenolide poison：カルデノリド毒
cardiac glycoside：強心配糖体
cardiac glycoside insensitivity：強心配糖

体に対する非感受性
Cardinium：カルディニウム
carina：(複 .-nae)、稜線、隆起線
caring for young：育仔(いくし)
carnivore：肉食(性)哺乳類、肉食性動物
carnivorous：肉食性の
carotenoid：カロチノイド
carrion：腐肉、死肉
carrying capacity：環境収容力
case：案件、格、事例
caste：カースト(階級制度)
casual release：思いつきの放蝶
catabolism：異化
catalyze：引き起こす、接触作用を及ぼす
category--subcategory ratio：カテゴリ - サブカテゴリの比、分類項目 - 細分類項目の比
category--subcategory taxonomic ratio：カテゴリ - サブカテゴリの分類数比
catenate flight：連結飛行、連結飛翔
caterpillar：幼虫(蝶／蛾の幼虫)、イモムシ、ケムシ
caterpillar stage：幼虫期
cation：カチオン、陽イオン
caudal：後端にある、尾(状)の
causal factor：成立要因
causative agent：原因病原体
cautery：焼灼法(しょうしゃくほう)
CBD：生物多様性条約(Convention on Biological Diversity)
CCYV：ウリ類退緑黄化ウイルス、ウリ類退緑黄化病(Cucurbit Clorotic Yellows Virus)
cDNA：相補的DNA(complementary DNA)
cease：中止する、停止する

cecropia moth：セクロピアサン
cecropin：セクロピン
celestial compass：天体コンパス
cell：(翅)室、中室、細胞
cell autonomous：細胞自律的
cell bar：中室の棒状紋
cell body：細胞体
cell count：細胞計数
cell end：中室端斑
cell homogenate：細胞磨砕物
cell image：細胞像
cell interaction：細胞間相互作用、細胞相互作用
cell-autonomous manner：細胞自律的な方法
cellular arrangement：細胞の配列
cellular component：細胞成分
cellular pattern formation mechanism：細胞の配列パターン形成機構
cellulose：セルロース
Cenozoic era：新生代(地質時代の一時代)
census：調査を行う、調査、センサス
census datum：(複 .-ta)、調査データ
center of distribution：分布の中心
center of frequency：頻度の中心
center of maximum variation：最大変異の中心
center of occurrence：発生の中心
center of origin：起源の中心
center of speciation：分化の中心地、分化の中心
centiMorgan：センチモルガン(遺伝学的な距離の単位)
central：中央部
central band：中央部の縞状バンド

Central China：中国中部
central focus：中心フォーカス
central fusion：中央融合型
central nervous system：中枢神経系
central neurosecretion cell：中央神経分泌細胞
central signalling region：シグナル伝達の中心部
central silk girdle：中央の帯糸
cercus：尾角、尾毛
cervical sclerite：頸節片
cervical shield：頸部シールド
cessation：休止、中断
cf.：「ラテン語の confer の略号で,『参照する』,『比較する』といった意味を持つ記号」,「『おそらくこの種であろう』と推測される場合,『属名 cf. 種名』の形式で表記」
CFC：隠れた雌による選り好み、雌による密かな性選択（Cryptic Female Choice）
CG：強心配糖体（Cardiac Glycoside）
chalaza：カラザ（卵の中で卵黄を安定させるひも）
challenging question：挑戦的な質問
change of photoperiod：日周変化、光周期変化、日長変化
chaparral：チャパラル（米南西部の矮性カシの木の密林）、イバラのやぶ
Chapter：章
character：形質、性質、特性、特徴
character displacement：形質置換
character map：キャラクターマップ
character mapping：キャラクターマッピング
characteristic：特性

characteristic distribution：特徴的分布
characteristic protective coloration：特徴的な保護色
charaxine butterfly：フタオチョウ
charcoal：木炭
chase：追飛行動
chasing：追跡飛翔、追飛
chauvinism：偏重主義、対外強硬主義
cheater：チーター
checkered pattern：格子縞、まだら模様
checkerspot butterfly：ヒョウモンモドキ
chemical：化学物質、化学薬品
chemical control：化学的防除
chemical cue：化学刺激、化学的な刺激
chemical ecology：化学生態学、ケミカルエコロジー
chemical halo：ケミカルハロー、化学的な暈（かさ）
chemical mimicry：化学擬態
chemical sense：化学の感覚、化学感覚
chemical structure：化学構造
chemically mediated signal：化学的媒介による信号
chemo-：化学 -〔ギリシャ語〕
chemoreception：化学受容
chemoreceptor：化学受容器、化学受容体
chemosensillum：化学感覚子
chemosensory gene：化学的感覚遺伝子
chemosensory hair：化学感覚毛
chemosensory neuron：化学感覚ニューロン、化学感覚神経細胞
chemosensory protein：化学感覚タンパク質
chemosensory regulation of feeding：化学感覚による摂食行動制御
chemotactile receptor：接触化学受容器

chestnut band：クリ帯
chevron：山形紋
chickadee：チッカディ（シジュウカラの仲間の小鳴鳥）
chilling：低温化、冷たい
chilling period：冷蔵期間
chimera：キメラ（同一個体中に遺伝子型の違う組織が互いに接触して存在する現象）
China mountain：中国山地
Chinese type：中国型
chitin：キチン（質）
chitinous ring：キチン環
cholesterol：コレステロール
chordata：脊索動物門
chordate：脊索動物
chordotonal organ：弦音器官
chordotonal sensory organ：弦音感覚器官
chorion：卵殻
chrom-：色 -〔ギリシャ語〕
chromatic adaptation：色彩適応
chromatogram：クロマトグラム
chromo-：色 -〔ギリシャ語〕
chromophore-binding pocket：発色結合ポケット
chromosomal evolution：染色体進化
chromosomal fusion：染色体融合
chromosomal map：染色体地図
chromosomal organization：染色体の構成
chromosome：染色体
chromosome level：染色体レベル
chronic：慢性の
chrono-：時 -〔ギリシャ語〕
chronobiology：時間生物学
chronological order：時代的順序
chronology：年代（推定学）、年表

chrysalis：蛹（さなぎ）、クリサリス（ギリシャ語で「金」の意味）
CI：細胞質不和合（Cytoplasmic Incompatibility）
Ci：肘脈中断遺伝子（翅脈が途切れている）（gene Cubitus-interruptus）
CI, 95%：95% 信頼区間（95% Confidence Interval）
cicada：セミ
ciliary opsin：繊毛型オプシン
ciliary photoreceptor cell：繊毛型光受容細胞
ciliary-type：繊毛型
cilium：（複 . -ia）、繊毛〔ラテン語〕
circadian：概日性、サーカディアン
circadian clock：概日時計
circadian hypothesis：概日仮説
circadian oscillation：概日振動
circadian rhythm：概日リズム、日周リズム
circled region：丸で囲った領域
circular flight：旋回飛翔、回転飛翔
circular flight area：旋回飛翔領域、旋回飛翔圏
circular-flying individual：旋回飛翔個体
circulatory system：循環系
Circum Japan Sea Area：周日本海地域
circum-：周囲 -、周辺 -〔ラテン語〕
circumpolar：極周辺の、極域周辺の
circumpolar species：環北極種
circumstantial line of evidence：一連の情況証拠、一連の状況証拠
cis regulatory element：シス調節要素、シス調節エレメント
CITEs：絶滅のおそれのある野生動植物の種の国際取引に関する条約、ワシントン条約（Convention on

International Trade in Endangered species of Wild Fauna and Flora)
citrus swallowtail butterfly：アフリカオナシアゲハ、オナシアゲハ
clade：分岐群、クレード、単系列、完系列、単系統
clade selection：分岐選択
cladistic analysis：分岐解析
cladistics：分岐論、分岐分類主義、分岐学
cladogenesis：分岐進化、クラドジェネシス
cladogram：分岐図、分岐関係図
clasper：把握器(はあくき)、クラスパー、交尾器
class：綱(分類階級の「こう」)、級(【統計学】)
classic biogeography：古典的の生物地理学
classical biological control：伝統的の生物的防除
classical hybrid inviability：古典的な雑種不和合性
classification：分類、類別
classification unit：分類単位
classified object：分類された対象群
claw：爪
clay treatment：粘土処理
clear asymptote：明らかな漸近線、完全な漸近線
clearing：開拓地、空地
clearing population：開拓地に適応した個体群
Clearwing Butterflies：トンボマダラ亜科
cleft of rock：岩の裂け目
cliff：崖(がけ)、絶壁
cliff-dwelling：崖を住居とする
climate：気候

climate change：気候変動
climate difference：気候的差異
climate regulation：気候調節
climate warming：気候の温暖化
climatic adaptation：気候適応
climatic condition：気候条件
climatic cooling：気候の冷涼化
climatic datum：(複．-ta)、気候データ
climatic lethal limit：気候的な致死限界
climatic zone：気候帯
climax：極相
clinal：クライン的
clinal variation：クライン的変異
cline：クライン、連続変異、勾配、傾斜
clone：クローン、栄養分枝系
close proximity on the chromosome：染色体上で近接した
close relation：近縁
close relative：近縁種
close-up：クローズアップ、大写し
close-up cross-section：拡大断面図
closed population：閉鎖個体群
closely related species：近縁種
closely related sympatric pair：近縁で同所的に生育するペア
closely related taxon：近縁な分類群
closest relative：最近縁種
club：棍棒、棍棒状部、触角先端部、クラブ、触角の先端の太い棒
cluster：クラスター
cluster structure：クラスター構造
clutch：卵塊
clypeus：(複．-pei)、唇基部、頭楯(とうじゅん)、頭楯板、頭盾〔ラテン語〕
cM：センチモルガン(遺伝学的な距離の単位)(centiMorgan)

co- : 共同 -、相互 -、共通 -〔ラテン語〕
co-author : 共著者、共同執筆者
co-mimic : 共通擬態種、相互擬態種
co-worker : 共同研究者
coastal : 沿岸の
coastal area : 海岸域
coastal vein : 前縁脈(ぜんえんみゃく)、C 脈
coat : 覆う
coauthorship : 共著者、共同執筆者
cobalt lysine : コバルトリジン
cobalt-coloring method of axon : 軸索のコバルト染色法
cocoon : 繭(まゆ)
cocoonase : コクナーゼ
code : 規約、暗号、コード
coding region : コード領域、コードしている領域
codominant molecular marker : 共優性分子マーカー
codon : コドン
codon-based maximum-likelihood analysis : コドンに基づく最尤解析
coecum penis (coe) : 陰茎盲のう
coevolution : 共進化
coexist sympatrically : 同所的に共存する
coexistence : 共存
coexisting strain : 共存系統
cohort : 区（補助的な分類階級の「く」）、同時出生集団、コホート、同類
cohort analysis : コホート解析
COI : チトクロム酸化酵素サブユニット 1（Cytochrome Oxidase subunit 1）
coinfect : 重複感染する、共感染する
coinfecting Spiroplasma strain : 重複感染しているスピロプラズマ系統
cold acclimatization : 寒冷順化

cold blooded : 変温の
cold chain : 低温系列
cold hardiness : 耐寒性
cold paralysis : 寒冷麻痺(かんれいまひ)
cold patroller : 体温の低い巡回者、低体温巡回型
cold region : 寒帯地域
cold shock : 低温ショック、寒冷ショック
cold-hardiness : 耐寒性
cold-season phenotype : 寒候期の表現型
Coleman method : コールマン法
Coleoptera : コウチュウ目、鞘翅目
collapse : 衰弱する
collar : えり、カラー
collate : 集めて分析する
collect : 採集する、収集する
collecting pressure : 採集圧
collection : コレクション
collection efficiency : 採集効率
collection point : 採集地
collective group : 寄集群
collector : 収集家、コレクター
colleterial gland : 粘液腺
colonisation : 移住
colonisation bottleneck : 定着ボトルネック
colonisation distance : 繁殖地間の距離
colonised site : 集団繁殖地
colonising history : 定着の歴史
colonist : コロニスト、定着者
colonization : 移住、移入、入植、定着
colonization ability : 定着力
colonization event : 定着イベント
colonization probability : 定着確率
colonization rate : 定着率、移入率
colonizing ability : 定着力、定着能力、移住能力

colonizing population：定着しつつある個体群

colony：集団繁殖地、コロニー

color：色彩

color band：色帯

color constancy：色（彩）恒常性

color field：カラーフィールド

color form：色彩型

color information：色彩情報

color marking：色斑

color pattern：色彩パターン、色彩斑紋

color pattern diversification：カラーパターンの多様性

color pattern shift：カラーパターン変動、カラーパターン変化

color polymorphism：色彩多型

color preference：色選好性

color preference map：色彩選好地図

color tone：色調

color variation：色彩のバリエーション

color vision：色覚

color-opponent response：色対比型応答、色対立型応答、反対色反応

color-pattern region：カラーパターン領域

coloration：色彩化、着色

Colore naturali edita - Lepidoptera：原色蝶類検索図鑑〔ラテン語〕

colore naturali edita：原色版、天然色版〔ラテン語〕

colored illustration：色図版、カラー図解

Colored illustrations of the butterflies of Japan：原色日本蝶類図鑑

Colored illustrations of the insects of Japan：原色日本昆虫図鑑

colored lateral filtering pigment：外側の着色フィルタリング色素

coloribus naturalibus：原色〔ラテン語〕

columnar neuron：カラムを構成しているニューロン、円柱のニューロン

combination：結合

combination of "hot-cold" and "dry-wet"：寒暖と乾湿の組合せ

combination of mate choice experiment：配偶者選択の組合せ実験

combinatorial probability：組合せ確率

combinatoric theory：組合せ（理）論

combined action：組合せ操作

Comet：縞模様

comma-shaped red marking：赤い半月紋

commensal：片利共生的

commensal interaction：片利共生的相互作用

commensalism：片利共生（へんりきょうせい）

Commission：審議会

commitment：コミットメント

common ancestor：同一祖先種、共通祖先種、共通祖先

common ancestor of primate：霊長類の共通祖先

common ancestral group：共通の祖先群

common blue butterfly：ウスルリシジミ、イカロスシジミ

common blue morpho butterfly：ペレイデスモルフォ

common bluebottle butterfly：アオスジアゲハ

common grackle：オオクロムクドリモドキ

common imperial blue butterfly：エバゴラスヒスイシジミ

common name：通称名、俗名

common opal butterfly：アフリカサバクシジミ

common phenomenon：共通の現象
common species：普通種
common, diurnally active, mute butterfly：通常、昼行性で音を出さない蝶
common-environment rearing：共通環境での飼育
communal foraging behavior：共同採餌行動
community：群集
community diversity：群集多様性
community ecology：群集生態学
community ecology study：群集生態学的研究
community module：群集モジュール
community structure：群集構造
community/ecosystem genetics：群集・生態系遺伝学
comparable density：比較可能な密度
comparative mapping：比較マッピング
comparative morphological method：比較形態学的方法
comparative study：比較研究
compartment：区画、コンパートメント、領域
compartmentation：構造化、区画化
compensatory growth：補償成長
competition：競争
competition theory：競争理論、競合説
competitive interaction：競争的相互作用
competitor：競争種
competitor relationship：競争者との関係
complementary DNA：相補的DNA
complementary sex determination：相補的性決定
complete darkness：暗黒
complete metamorphosis：完全変態

completed adult body formation：完全な成体形成
completed adult morphogenesis：完全な成虫の形態形成
complex：複合体
complex of gene：遺伝子複合体
complex peak：複合ピーク
complex statistical correction：複雑な統計的補正
component：構成分子、構成要素
component species：構成種
composite interval mapping：複合区間マッピング法
compound：複合（の）
compound action potential：複合活動電位
compound eye：複眼
compound microscope：複合顕微鏡、複式顕微鏡
compound name：複合名
computationally intensive：計算集約的、計算性を強化したもの
computer simulation：コンピュータシミュレーション
Comstock-Needham system：コムストック‐ニードハム（ニーダム）の体系
concave：くぼんだ、凹む
conceal：隠す
conceivably：ひょっとすると、おそらく
concentration：濃度
concentric circle：同心円
concentric eyespot pattern：同心円状の眼状紋パターン
concentric ring：同心円状の環
conclusion：結論
conclusive prediction：決定的な予測
concordant：調和した、一致した

concordant pattern：一致したパターン
condensed sex chromatin body：凝縮性染色質体
conditional：条件付きの
conditional proposal：条件付きの提案(書)
cone：円錐形
cone photopigment：錐体視物質
cone photoreceptor cell：錐体視細胞
cone tweeter：円錐ツイーター
confer：与える
confidence interval：信頼区間
confidence limit：信頼限界
confirm：確かめる、裏付ける
confound：混同する
confronting behavior：対峙する行動
congeneric：同属の
congeneric coexistence：同属の共存
congeneric species：同属種
congregate：集まる
conical morphogen gradient：モルフォゲンの円錐形濃度勾配
conifer：針葉樹
connected population：孤立化していない個体群
connecting vowel：結合母音
consecutive in time：時間に連続した
consensus sequence：コンセンサス配列、共通配列
consent：同意、承諾、許可
conservation：保護、保存、保全
conservation biogeography：保全生物地理学
conservation biology：保全生物学
conservation decision：保全決定
conservation ecology：保全生態学
conservation effort：保護努力、保全努力

conservation expenditure：保護支出、保護費用
conservation introduction：保全的導入
conservation organization：保護団体、保護機関
conservation strategy：保全戦略
conservationist：自然保護者、環境保全主義者
conservative hypothesis：保守的仮説
conservatively：保守的に
conserve：保全する
conserved developmental pathway：保存された発生経路
conserved name：保全名
conserved PCR primer：保存されたPCRプライマー
conserved polymerase chain reaction primer：保存されたポリメラーゼ連鎖反応用プライマー
conserved supergene locus：保存されたスーパー遺伝子座
conserved work：保全された著作物
consistent amplification：一貫した増幅
conspecific：同種の、同種類、同じ種類
conspecific communication：同種間コミュニケーション、同種間通信
conspecific female：同種の雌
conspecific individual：同種の個体
conspecific mating：同種間の交配
conspecific phenotype：同種の表現型、同種表現型
conspecific recognition：同種認知
conspecific wing：同種の翅
conspicuous coloration：目立つ色、顕著な色
conspicuous eyespot：目立つ眼状紋

conspicuous ventral wing marking：腹側の翅の目立つ斑紋
conspicuousness：目立つこと、顕著さ
constant temperature：恒温
constant temperature regime：恒温条件、一定温度条件
Constitution：審議会規則、制定、制度、法令
consumer：消費者
consumption of fat：脂肪消費
consumptive effort：消費効果
contact chemoreceptor：接触化学感覚器、接触化学受容器
contact pheromone：接触フェロモン
contact zone：接触帯、接触域
containment：封じ込め
containment area：封じ込め地域
contig：コンティグ、整列群、連結断片
continent：陸地
continental：大陸産
continental characteristic climate：内陸性気候
contingency plan：緊急時対応計画
continuous distribution：連続分布
continuous expenditure：継続的な支出
continuous flight behavior：連続飛翔行動
continuous nucleotide：連続した塩基、連続したヌクレオチド
continuous parental antibiotic treatment：親の継続的な抗生物質処理
continuous ring：連続したリング
continuously flying type：連続飛翔型
contradict each other：相矛盾する
contradiction：反論
contrary effect：逆効果
contrast：コントラスト

contribution ratio：寄与率
control：防除、制御、管理、対照、統制
control measure：防除措置
control mechanism：調節機構
control threshold：要防除密度
controlling factor：支配要因、決定要因
controlling mechanism：調節機構、制御機構
controversial：議論の余地がある
controversy：論争
Convention on Biological Diversity：(CBD)、生物多様性条約
Convention on International Trade in Endangered species of Wild Fauna and Flora：(CITEs)、絶滅のおそれのある野生動植物の種の国際取引に関する条約
Convention on International Trade in Endangered species of Wild Fauna and Flora Convention：(CITEs)、ワシントン条約
conventional sodium dodecyl sulfate-proteinase K digestion：従来のデシル硫酸ナトリウム - プロテイナーゼ K による消化
converge：収斂する、収束する
convergence：収斂(しゅうれん)、収斂現象
convergent amino acid change：アミノ酸の収束的変化
convergent change analysis：収束的変化解析
convergent evolution：収斂した進化、収斂進化、収束進化
convergent functional-site evolution：機能的部位の収斂進化
convergent mimicry：収斂的擬態

convergent phenotype：収斂した表現型、集束した表現型
convergently：収斂的に
convex bubble：凸状の泡、凸状の気泡
convex inner membrane：凸状の内膜
convincing example：確証例、確信させてくれる例
cool place：冷涼な場所
cool temperature：低温、冷温
cool temperature treatment：低温処理
cool- and warm-temperate zone：冷温帯域
cool-temperate deciduous broadleaf forest：冷温帯性落葉広葉樹林
cool-temperate deciduous forest：冷温帯性落葉樹林
cool-temperate habit：冷温帯性
cool-temperate species：冷温帯性の種
cool-temperate zone：冷温帯
cooled pupa：冷却蛹
cooling：寒冷化
cooling treatment：冷却処理
coordinated evolution：同調した進化
Coppers：シジミチョウ亜科
coppice：雑木林
coppicing：雑木林の伐採、コピス
copulated individual：交尾個体
copulation：交尾（こうび）
copulatory：交尾の
copulatory aperture：交尾孔
coral reef fish：サンゴ礁に生息する魚、サンゴ礁魚
core area：活動中心部
coremata：発香総
cornea：角膜
corneal lens：角膜レンズ
corner：かど（角）、辺縁部

cornutus (cor)：（複 . -ti)、射精のう棘〔ラテン語〕
corpus allatum：（複 . -ta)、アラタ体〔ラテン語〕
corpus bursa：交尾のう
corpus cardiacum：（複 . -ca)、側心体〔ラテン語〕
correct original spelling：正しい原綴（つづ）り
correlate：相関がある
correlation：相関、相関関係
corresponding author：連絡著者、連絡先となる著者、コレスポンディングオーサー、通信担当者、責任著者
correspondingly large and significant peak：対応する大きく有意なピーク
corridor：コリドー、回廊
corrigendum：（複 . -da)、正誤表
cosegregation：共分離
cosmopolitan distribution：世界各地の分布
cosmopolitan species：普遍種
cosmopolite：普遍種
cost：出費、コスト、損失、費用
costa：前縁、前縁脈、C脈〔ラテン語〕
costal：前縁部、前縁室
costal fold：前縁脈のひだ、前縁褶
costal margin：前縁
costal vein：前縁脈、C脈
cotype：コタイプ、副基準標本
coumarin：クマリン
counter behavior：対立行動
counteracting power：消去力
counterpart：代置種
countless site：無数の場所
countryside：田園地帯

countryside butterfly：里山の蝶
court：求愛する、誘う
courtesy：好意
courtship：求愛行動
courtship behavior：求愛行動
courtship dance：求愛ダンス
courtship display：求愛誇示
courtship element：求愛の行動要素
courtship pheromone receptor：求愛フェロモン受容体
courtship preference：求愛選好
covariate：共変量、共変数、共分散分析
cover scale：カバースケール
cover site：隠れ場所
coverage：包括度、カバー率
coverslipped：覆った
coxa：（複 . -ae）、基節(きせつ)〔ラテン語〕
coyness：はにかみ、内気
cracker butterfly：カスリタテハ属の蝶
crackling sound：パチパチという音
Cramer's blue morpho：レテノールモルフォ
cranial：前(ぜん)
craw：爪(つめ)
cream：乳白色、クリーム
creased appearance：ひだ模様
creature：生物、創造物
credibility value：信頼性の値
cremaster：懸垂器、尾鉤
crenulate：小鈍鋸歯状の
crepuscular：薄暮活動性、薄明、薄暮時の
crescent-shaped：三日月形の
Cretaceous period：白亜紀
crevice：割れ目
cricket：コオロギ
criminal offence：犯罪行為

crimson tip butterfly：ダナエツマアカシロチョウ
critical day-length：臨界日長
critical illuminance：限界照度
critical light intensity：限界光量
critical mate preference cue：配偶者選好の臨界刺激
critical night-length：臨界夜長
critical number of day：臨界日数
critical photoperiod：臨界日長、臨界光周期
critical species：危篤種
critical threshold：臨界閾(いき)値
critical value：臨界値
critically endangered native species：危機的な在来の絶滅危惧種
crocale form：（ウスキシロチョウの）無紋型、クロカレ型
crochet：鉤爪(かぎづめ)、鉤爪
crop：素のう
cross：交雑種、交雑、断面、クロス、交配
cross even twice：さらに二回交雑する
cross habituation：交差慣化
cross once：一回交雑する
cross rib：横梁
cross section：断面、断面図
cross-band：網目模様
cross-hatched area：格子囲い領域、クロスハッチング領域
crossbreed：雑種、異種交配種
crossing：種間交雑
crossing scheme：掛け合わせ様式、異種交配様式
crossing-over rate：交叉率(こうさりつ)、交差率
crossvein：横脈

crosswind drift：横風ドリフト、横風偏流
crowding：群がり、込み合い、群集
crowding condition：混み合う条件、混雑条件
crucial stage in speciation：種分化の決定的な段階
crumpled wing：しわしわの翅
cryophase：低温期
cryoprotectant：冷凍保存、低温障害防御物質
crypsis：隠蔽(いんぺい)
cryptic color：隠蔽色、保護色
cryptic coloration：保護色、隠蔽色
cryptic female choice：隠れた雌による選り好み、雌による密かな性選択
cryptic mimicry：隠蔽型擬態
cryptic species：隠蔽種
crypticity：隠蔽性、隠蔽色
cryptobiosis：乾燥休眠、クリプトビオシス(「隠された生命活動」の意味)
crystal formation：結晶形成
crystalline cone：円錐晶体
CSD：相補的性決定（Complementary Sex Determination）
CSP：化学感覚タンパク質（ChemoSensory Protein）
CT：要防除密度（Control Threshold）
Cu：肘脈(ちゅうみゃく)（Cubitus）
cubital：肘脈の
cubital crossvein：肘横脈
cubital vein：肘脈、Cu脈
cubitus：肘脈、Cu脈
cubitus interruptus：キュビタスインターラプタス（転写制御因子）、肘脈中断
cucullus：交尾弁端部〔ラテン語〕
cultivate：開墾する、養殖する、栽培する

cultivated land：耕作地、耕地
cultivation：耕作、開墾
cultural control：耕種的防除
culture medium：培養基、培地
cultured cell：培養細胞
cumulative heat unit：積算温量
cumulative number of species：累積種数
cumulative temperature：積算温度
curled：曲がった
curve-fitting extrapolation method：曲線当てはめ型外挿法、カーブフィッティング外挿法
curved wing-tip：鉤状の前翅先端
cut flower：切り花
cut *Lantana* sp. flower：ランタナ種の切り花
cuticle：クチクラ、表皮、キューティクル
cuticle scale：表皮鱗粉
cuticular hydrocarbon：体表炭化水素
cuticular surface：皮膚表面
cuticular ultrastructure：表皮の超微細構造
cutting of dry leaf：枯れ葉の切断
cv.：園芸品種（cultivar）、栽培変種
cyanide release：シアン化物の放出
cyanogenesis：シアン生成、青酸生成能
cyanoglucoside：シアノグルコシド
cyclic cicada：周期ゼミ
cyclops：単眼体
Cydno longwing butterfly：シロオビドクチョウ
cysteine arrangement pattern：システイン配置パターン
cytochrome：チトクロ（ー）ム
cytochrome oxidase：チトクロム酸化酵素
cytochrome oxidase subunit 1：チトクロム酸化酵素サブユニット1

cytochrome oxidases I and II：チトクロームオキシターゼ I と II
cytochrome P450：チトクローム P450
cytological observation：細胞学的観察
cytologically：細胞学的に
cytoplasmic incompatibility：細胞質不和合
cytoplasmic incompatibility microorganism：細胞質不和合微生物
cytoplasmic sex ratio distorter：細胞質因子による性比歪曲因子、細胞質性比歪曲因子
cytoplasmically inherited gene：細胞質遺伝性遺伝子

d

D：暗期（Dark）、表面（Dorsal）、上面、背側
D-statistic：D 統計量
d.f.：自由度（degree of freedom）
dagger-like：短刀状の
daily activity：日周活動、日周行動
daily periodicity：日周サイクル
damaged area：損傷領域
damaging invasive species：有害な侵入種
damp area：湿地
danaid butterfly：タテハチョウ科の蝶
danaidone：ダナイドン（性行動刺激物質で、雄のヘアペンシル成分／性フェロモン成分）
dark box：暗箱
dark brownish color：こげ茶色
dark form：暗色型、暗化型
dark mimic：暗化擬態
dark period：暗期
dark phase：暗期、暗相
dark reddish-brown color：暗赤褐色
dark time：暗期
dark treatment：暗黒処理
dark-adapted live insect：暗順応性の生きた昆虫
dark-time measurement：暗期測定
darken light wavelength：減光波長
darker, shadier place：より暗く、より薄暗い場所
darkness：暗所
darting flight：矢のようにすばやく飛ぶ
Darwin：ダーウィン
Darwin's finch：ダーウィンフィンチ（キンパラ科の鳥）
data logger：データロガー、データ計測器
database：データベース
date：日付
date of publication：公表の日付、刊行日
daughter cell：娘細胞、嬢細胞
dawn：薄明、夜明
day degree：日度
day length：日長
day-flying：昼行性の
day-length：日長、日長時間
day-old：日齢
daylight hour：日照時間
dazzle：目がくらむ
dB：デシベル（decibel）
DD：全暗黒、恒暗（continuous Dark あるいは Dark-Dark）
DDC：ドーパ脱炭酸酵素、ドーパデカルボキシラーゼ（DopaDeCarboxylase）
de novo：初めから、新たに〔ラテン語〕
de-：下降 -、否定 -〔ラテン語〕
de-efferented：神経遮断の
dead feigning：擬死

dead insect : 死んだ昆虫
dead leaf : 枯れ葉
dead mimicry : 擬死
decapentaplegic : デカペンタプレジック遺伝子(シュウジョウバエの形態形成遺伝子)
decapitation : 断頭
decaying : 腐敗しかけた、朽ちる
decent rule of thumb : 厳しくない経験則
deceptive flower : だまし花
deciduous : 落葉性の
deciduous broadleaf forest : 落葉広葉樹林
deciduousness : 落葉性
decision making : 意志決定、意思決定
decisive step : 決定的なステップ
deck of shuffled card : シャッフルされたカードの束
Declaration : 布告書
decline : 衰亡
decline of genetic diversity : 遺伝的多様性の低下、遺伝的多様性の衰亡
decline stage : 滅亡の段階
declining butterfly : 衰亡しつつある蝶
decomposer : 分解生物、分解者
decomposition : 分解、腐敗、腐朽
decorate : 飾る
decrease : 低下する、減少する
decreasing abundance : 減少する存在量
deem : 見なす
deep : 深(しん)、深みのある
deep-red in upperside : 小豆色の地色
deer : シカ
defending behavior : 防御行動
defense activity : 防衛行動
defense compound : 自己防衛物質
defensin : ディフェンシン

defensive regurgitation : 防御的吐き戻し
defensive substance : 防御物質
definition : 定義
deflect : そらす、曲折させる
deflection of predator attack : 捕食者攻撃のぶれ
deforestation : 森林破壊、森林伐採
deformed testis : 奇形の精巣
deformed wing : 奇形の翅
degenerate : 退化する、悪化する
degenerate forward primer : 順方向縮重プライマー、縮重フォワードプライマー
degenerate primer : 縮重プライマー
degenerative : 退化的
degradation : 低下
degree of change : 変化程度
degree of crowding : こみあいの程度、混み具合
degree of occurrence : 出現程度
degree-day : 日度
dehydrate : 脱水する
dehydration : 脱水
delayed larva : 遅れた幼虫、遅発幼虫
deleterious allele : 有害対立遺伝子
deleterious effect : 有害作用、有害効果
deleterious genetic effect : 有害な遺伝的影響
deleterious recessive : 有害な劣性種
deletion : 欠失
deliberate introduction : 故意の導入
deliberate release of butterfly : 故意の放蝶
delineation : 境界
deme : 地域集団、ディーム
demographic character of metapopulation : メタ個体群の人口学的特性

demographic concern：人口学的関心

demographic inertia：人口学的慣性

demographic parameter：人口学的パラメータ、人口学的母数

demographic scenario：人口学的筋書き、人口学的シナリオ

demographic stochasticity：人口学的確率性

demographically：人口統計的に

dendrite：樹状突起

dendritic tip：樹状突起先端部

dengue fever：デング熱

dense：稠密（ちゅうみつ）な、濃密な

dense genus：稠密な属

dense patch：密集したパッチ

dense species：稠密な種

dense zone：濃密な地帯

dense, shady wood：密集した日陰の多い森

densely distributed marker：密に分散されたマーカー

denser：より濃い

density effect：密度効果

density of microtrich：微毛密度、微毛の密度

density-dependent dispersal：密度依存的分散

density-dependent process：密度に依存する過程

density-independent process：密度に依存しない過程

density-mediated indirect effect：密度の変化を介する間接効果

dentate：歯のある

deoxyribonucleic acid：デオキシリボ核酸（DNA）

dependent pattern：依存パターン、翅脈依存パターン

deploy：配置に付かせる

depression：衰弱

derived group：派生グループ

dermal gland：皮膚腺

Dermaptera：ハサミムシ目、革翅目

descending flight：相手を下方に抑え込もうとする飛翔行動

describe：記載する

description：記載

desert：砂漠、荒原

desiccation：乾燥

designate：指定する

designation：指定

detectable association：検出可能な関連

deteriorate：衰える、悪化する、劣化する

deteriorated food condition：悪化した食物条件

deterioration：悪化

determinant factor：決定要因

determination：決定機構

deterrent：抑止力、抑制因子

detoxication enzyme：解毒酵素

detritophagy：腐泥食性、腐食性

detrivore：腐泥食（性）生物、腐食（性）生物

devastating attack：壊滅的な打撃

developed country：先進国

developing forewing：前翅の発生

developing leg：脚発生

developing wing：翅発生

development：発生、発育、成長

development construction of～：～の造成、～開発

development history：形成史、発展史

development induction：発育誘導
development phase：発育相
development program：発育プログラム
development rate：発育速度、生長速度、発育率、発生率、成長率、発達率
development stage：発育相、発育段階
development zero：発育零点
developmental biology：発生生物学
developmental delay：発育遅延、成長遅延
developmental endemism：発展的固有
developmental gene：発生遺伝子
developmental genetic mechanism：発生遺伝機構
developmental homeostasis：発育安定性、発生的恒常性、発育恒常性
developmental homologue：発生ホモロジー、発生相同
developmental mechanism：発生機構
developmental pathway：発生経路
developmental process：発育過程
developmental season：発育季節
developmental stage：発育期、発育ステージ
developmental switching mechanism：発生上のスイッチ機構
developmental time：発育時間、発達時間
developmental zero point：発育零点
developmentally arrested pupa：発育が抑制された蛹、発育が阻害された蛹
devote：充てる
DH：休眠ホルモン（Diapause Hormone）
di-：2-、二-、双-〔ギリシャ語〕
dia-：通して-、間の-〔ギリシャ語〕
diachronicity：通時性、経時性（単系統種）
diacritic mark：区別的発音符
diagnosis：診断

diagnostic PCR detection：診断 PCR 検出法、診断 PCR 法による検知
diagonal bar：斜めの条線、斜め線
diagonal broken line：対角破線
diamond pattern：ひし形
Diana butterfly：ダイアナヒョウモン
diapause：休眠（主に動物の）、発生休止状態の休眠、休眠期
diapause ability：休眠能力
diapause characteristic：休眠特性
diapause condition：休眠条件
diapause destined larval period：休眠になる場合の幼虫発育期間、休眠性幼虫発育期間
diapause development：休眠成長、休眠発育
diapause egg：休眠卵
diapause entry：休眠突入
diapause factor：休眠因子
diapause hormone：休眠ホルモン
diapause induction：休眠誘起
diapause induction ratio：休眠誘起率
diapause instar stage：休眠齢期
diapause intensity：休眠深度、休眠の深さ
diapause larva：休眠幼虫
diapause phase：休眠相
diapause program：休眠プログラム
diapause ratio：休眠率
diapause response：休眠反応
diapause site：休眠場所
diapause stage：休眠期
diapause state：休眠状態
diapause termination：休眠覚醒、休眠消去
diapause termination rate：休眠解除率
diapause-blocking effect：休眠阻止効果
diapause-destined larva：休眠予定幼虫、

休眠性幼虫
diapause-destined pupa：休眠予定蛹
diapause-inducing photoperiod：休眠誘起の日長、休眠誘起の光周期
diapause-maintaining mechanism：休眠維持機構
diapausing adult：休眠成虫、休眠型成虫
diapausing pupa：休眠蛹
diaphragm：隔膜、横隔膜
dichotomous key：二分岐のキー
dichotomy：二項対立、二分法、二分
dichromatic：二色型色覚の、二色性の
dicotyledon：双子葉植物
dideoxy terminator：ジデオキシターミネーター
diet：食料、飼料
diet quality：食料の質、餌の質
diet shift：食性転換
dietary nitrogen：食餌性窒素、食物の窒素
differance：差延（「空間的差異と時間的遅延」を意味する短縮語）〔フランス語の造語〕
difference in fitness：適応性の相違
difference of development stage：発育段階の差
different climatic region：異なる気候の地域
different genus：別属
differentiate：識別する、分化する
differentiation：分化
differentiation of food habit：食性分化
differently-pigmented scale：異なる色素で着色された鱗粉
difficult condition：悪条件
diffusion and freeze-back：拡散と寒冷押し戻し
diffusion process：放散過程、拡散過程
diffusion-reaction model：拡散反応系モデル
digestion：消化
digestive system：消化系
digital oscilloscope：デジタルオシロスコープ
digital signal processor：デジタル信号処理装置、デジタルシグナルプロセッサー
diglyceride：ジグリセリド
diglyceride-carrying protein：ジグリセリド運搬タンパク質
dilute：弱める、希釈する
diluted honey：薄めたハチミツ
diluted solution：希釈溶液
dilution：希釈液、希釈物
dilution effect：希釈効果
dim light：薄暗い光
dim monochromatic flash set：薄明かりの単色閃光セット
dimethylsulphoxide：ジメチルスルホキシド
diminish：減らす
dimorphism：二形性、二型
dip：（水平線の磁針の）伏角
diploid embryo：二倍体胚
Diptera：ハエ目、双翅目
dipteran insect：双翅目（そうしもく）の昆虫、ハエ目の昆虫
direct defense：直接防衛
direct driver：直接の影響要因
direct selection：直接的な選択
direct sequencing：直接シークエンス法
direct sunlight：直射日光

Direction：告示書
direction ratio：方向比
directional effect：方向性効果
directional flight：方向移動飛翔
directional flight ability：方向移動能力
directional selection：方向性選択、定方向性選択、定方向選択、方向性淘汰
dis-：分離 -、不 -、非 -、無 -〔ラテン語〕
disc：原基、（翅）円盤
disc equation：円盤方程式
discal：中央帯、中央部
discal cell：中室
discal lacuna：横脈くぼみ、横脈欠落部
discal spot：横脈紋
Disclaimer：棄権宣言
discontinuous distribution：不連続分布
discrete focus：分離した焦点
discrete wing pattern element：不連続な翅のパターン要素
discriminate：識別する、区別する
discrimination：識別
discussion：考察、議論
disease myxomatosis：疾患粘液腫病
disease outbreak：病気の大流行、病気の発生
diseased hybrid brood：病気の雑種同腹仔
disguise：擬態する
disinfection：消毒
disjunct distribution：隔離分布、分離分布
disorientate：方向感覚を失う
dispersal：分散、拡散、移動、放散
dispersal ability：分散力、分散能力
dispersal morph：分散型
dispersal polymorphism：分散多型
dispersal potential：分散の潜在能力
dispersal rate：分散率

dispersal tendency：分散性向、分散傾向
dispersal trait：分散形質
dispersant：分散剤
dispersed distribution：分散分布
dispersion ability：分散能力
display：誇示、表示
display behavior：誇示行動
disproportionate influence：より大きすぎる影響
disproportionately large：不釣り合いなほど多くの、かなり多くの
disprove：誤りを実証する、反証をあげる
dispute：反論する
disregarding abundance：無視できる程度の個体数
disrupt：崩壊させる、分裂させる
disruption：崩壊、分裂、分断
disruptive ecological selection：分断的生態選択、分断化生態選択、分断性生態選択
disruptive natural selection：分断自然選択、分断自然淘汰
disruptive selection：分断選択、分断淘汰
disruptive sexual selection：分断性選択、分断性淘汰
dissect：解剖する、詳細に調べる、切断する
dissected wing：切り取った翅
dissection：解剖体、解剖
dissection technique：解剖手法、解剖技法
dissimilation：異化
dissolution：崩壊
distal：末端の、末梢の、端部の、遠位
distal ablation：末端部位の切除
distal band：末端側の縞状バンド（外横線）
distal region：末端領域、末梢領域、遠位

領域
distal tip：外縁の先端、付属肢の先端
Distal-less expression：末端部のない遺伝子の発現
distal-less gene：末端部のない遺伝子
distal-less protein：末端部のないタンパク質
distally：末端側に、末梢側に
distant lineage：離れた系統
distantly related：遠縁の
distantly related co-mimetic species：遠縁の相互擬態種
distantly related species：遠縁種
distasteful：味の悪い、まずい
distilled water：蒸留水
distinct color pattern：明瞭なカラーパターン
distinct species：別種
distinct subspecies：別亜種
distinct west-east gradient：西から東にかけての明確な勾配
distinctive allele：明確に区別できる対立遺伝子
distort：ゆがめる
distortion：ゆがみ
distribute：分布する、分配する
distribution：分布
distribution change：分布の変化
distribution element：分布要素
distribution frequency：分布頻度
distribution of genus：属分布
distribution of organism：生物分布
distribution pattern：分布パターン、分布様相
distribution region：分布域
distribution route：分布経路

distribution type：分布型
distributional range of genus：属分布圏
distributional range of species：種分布圏
disturbance：攪乱（かくらん）
disturbance frequency：攪乱頻度
disturbance intensity：攪乱強度
disturbance treatment：攪乱処理
ditrysia：二門亜目
ditto underside：同裏面
ditto, UN：同裏面（ditto UNderside）
diuresis：利尿
diuretic hormone：利尿ホルモン
diurnal：昼行性の、昼間の
diurnal activity：日周活動、日周行動
diurnal behavior：日周活動、日周行動、昼行活動
diurnal butterfly：昼行性蝶
diurnal moth：昼行性蛾
diurnal periodicity：日周性
diurnal rhythm：日周リズム
diurnal wood nymph：昼行性の森林性若虫
diurnally active：日中に活動的な、日中に活動する、昼行性の
diurnally active and mute：昼行性で音を出さない
diverge：分岐する、分散する
diverged population：分岐した個体群
divergence：分岐、相違、分化、多様化
divergence between higher classification：高次分類間の分岐
divergence time：多様化期間、分岐時間
divergent clade：多様な分岐
divergent evolution：分岐進化
divergent function：異なる機能、多様な機能

divergent mate preference：多様な配偶者選好

divergent pattern：異なるパターン、分岐パターン

divergent wing pattern：多様な翅パターン

diverse age structure of tree：多様な樹齢構造

diverse array：分岐配列、多様な列挙

diverse variation：多様な変異

diversification：多様化

diversity：多様性、分岐

diversity index：多様性指標、多様性指数、多様度指数

divert：惑わせる

division：区分、区画、分裂

DmelGr43a：キイロショウジョウバエ（*Drosophila melanogaster*）の味覚受容体 43a

DMSO：ジメチルスルホキシド（Dimethylsulphoxide）

dN：非同義置換率（Non-synonymous substitution rate）

dN/dS：非同義・同義塩基置換数比（Non-synonymous/Synonymous nucleotide substitution ratio）

DNA：デオキシリボ核酸（DeoxyriboNucleic Acid）

DNA barcoding：DNA バーコーディング

DNA extraction：DNA 抽出

DNA marker assisted selection：DNA マーカー利用選抜

DNA polymorphism：DNA 多型

dodging：肩透かし、素早く逃げる、ジグザグ飛翔

domain：ドメイン（インターネット上で用いるコンピュータのグループ名や住所）、上界、変域、領域

dome-like shape：ドーム状の形

domestic alien species：国内由来の外来種、国内外来種

domestic animal：家畜

domestic native species：国内在来種

domesticated animal：家畜

dominance：優性、顕性

dominance of melanic element：黒化要素の優性

dominance relationship：優性関係

dominant：優性の、優勢形質、優占種、顕性

dominant mutation：優性突然変異

dominant white and recessive yellow allele：優性白色／劣性黄色の対立遺伝子

dominate and wide-distributed species：優勢な広域分布種

dominate species：優占種、優先種

donate：提供する、寄付する、寄贈する

donor：供与体、供与菌、ドナー、寄贈者

donor population：供与者個体群

donor site：提供側生息地

donor stock：ドナーストック

doom：運命づける

dopa：ドーパ（アミノ酸の一種）

dopadecarboxylase：ドーパ脱炭酸酵素、ドーパデカルボキシラーゼ

dopamine：ドーパミン（脳内の神経伝達物質）

dormancy：休眠（主に植物の）、活動休止状態の休眠

dormant phase：休止相

dorsal：背部の、背面の、上面の、表面の、背側(はいそく)、背(方)の

dorsal appendage：背側突起、背部の付属肢
dorsal basking：背面日光浴、開翅日光浴
dorsal direction：背面方向
dorsal dissection approach：背側解剖法
dorsal eyespot：背部眼状斑点、背眼状斑点
dorsal forewing：背側の前翅
dorsal lateral line：背側の側線
dorsal nectary organ：背部蜜腺
dorsal part of pars intercerebralis：脳間部背面
dorsal plane：背断面、背面図
dorsal retina：背側網膜
dorsal section：背断面、背面図
dorsal seta：背毛、背部の刺毛、背部の剛毛
dorsal side：背側
dorsal side of wing：翅背側
dorsal surface：背断面、背面図
dorsal surface of forewing：前翅背面
dorsal vessel：背脈管(はいみゃくかん)、背管
dorsal-basking：背面日光浴
dorsal-ventral axis：背腹軸、DV軸
dorsally：背側に
dorsoventral axis：背腹軸、DV軸
dorsoventral flight muscle：背腹方向に走る飛翔筋、背腹飛翔筋
dorsoventral flight musculature：背腹方向に走る飛翔筋組織
dorsum：後縁、後縁帯〔ラテン語〕
dorsum of forewing：前翅後縁
dot：小斑点
double-strain-infected all-female brood：二系統に感染したすべて雌の同腹仔、二重感染のすべて雌の同腹仔

double-stranded RNA：二本鎖RNA、二重鎖RNA
doublesex：両性
doubleton：2個体だけ現れた種、ダブルトン
doubling time：倍増時間
doubly infected female：重複感染した雌
doubly infected insect line：重複感染の昆虫系統
downgrade：優先度を下げる
downstream of sex determination cascade：性決定カスケードの下流
downstream of sex determination system：性決定システムの下流
downstream pathway：下流経路
downward selection：下方選択
Dpp：デカペンタプレジック遺伝子（*Decapentaplegic* gene）
drab little wing：茶褐色の小さな翅
draft genome assembly：ドラフトゲノムのアセンブリー
dragonfly：トンボ
drainage：乾燥化
draining：乾燥化
drawing：線画
dried specimen：乾燥標本
drift：浮動
drift selection：浮動選択
drink：吸水する
drinking：吸水中
drinking water：吸水
driver of change：変化の駆動因子
drizzle：霧雨
drooping tail：はかまの裾のような後翅（尾端）
drop of saline：塩水の滴下

droppings：糞
Drosophila：ショウジョウバエ属
drought：旱魃（かんばつ）、渇水、日照り、干ばつ
drumming：連打、ドラミング
dry climate：乾燥気候
dry forest：乾燥森林
dry forest region：乾燥森林地域、乾燥森林地帯
dry season form：乾季型、乾型
dry thorn-scrub habitat：乾燥有刺低木林の生息地
dry upland grass：乾燥高地の草、乾地草
dry zone：乾燥地帯
drying：乾燥、乾燥化
Ds：横脈紋（Discal spot）
dS：同義置換率（Synonymous substitution rate）
dsf.：乾季型（dry season form）
dsRNA：二本鎖RNA、二重鎖RNA（double-stranded RNA）
dual role：二つの役割
dull-colored：鈍い色の
dung：糞（動物の）
dung fly：クソバエ、糞蝿
duplicate gene：重複遺伝子
duplicate ultraviolet opsin：重複型紫外線オプシン
duration of larval instar：幼虫期の齢期間
duration of light exposure：照射される光の期間
duration of response：応答期間、反応継続期間
dusk：薄暮、薄暗がり、たそがれ
dust：散在する
dv：背腹方向に走る（dorsoventral）
DV axis：背腹軸、DV軸（Dorsal-Ventral axis）
dwell：棲息する
Dyar's law：ダイアーの法則
dynamic equilibrium：動的平衡
dynamic range：ダイナミックレンジ

e

E. coli：大腸菌（*Escherichia coli*の省略形）
E. coli colony：大腸菌コロニー
EAG：触角電図、触角電図法（ElectroAntennoGram）
eagle owl：ワシミミズク
ear：耳
ear drum：鼓膜
earliest stage of speciation：種分化の最初期
early Cenozoic era：新生代初頭
early detection：早期発見
early embryonic event：胚の初期発生イベント
early embryonic stage：胚発生の初期段階
early instar larva：若齢幼虫
early male-killing：初期型雄殺し
early morning：早朝
early phase of invasion：侵入の初期
early pupa：蛹初期
early pupal epidermis：蛹初期の表皮
early pupal wing：蛹初期の翅
early spring：初春、早春
early stage：幼生期
early warning of change：変化の早期警報
earnest：本気の、熱心な
earth：地球
earth's magnetic field：地磁気、地球の磁場

ecdysial droplet：脱皮顆粒
ecdysis：脱皮（だっぴ）、ぬけ殻
ecdysis triggering hormone：脱皮刺激ホルモン
ecdysone：エクジソン、エクダイソン（脱皮ホルモンの一種）
ecdysteroid：エクジステロイド（脱皮ホルモンの一種）
ecdysteroid hormone：エクジステロイドホルモン、脱皮ホルモン
ecdysteroid release：エクジステロイドの放出
ecdysteroid titer：脱皮ステロイド力価、エクジステロイド量
ecdysterone：エクジステロン（エクジソンに似た脱皮ホルモン）
eclose：羽化する
eclosed female：羽化した雌
eclosion：羽化、脱蛹
eclosion hormone：羽化ホルモン
eclosion rate：羽化率
eclosion season：羽化時期
eclosion termination：羽化消去
eco-chemical：生態化学
Ecocene：始新世
ecoevolutionary dynamics：生態進化ダイナミクス
ecological character：生態（学）的特性
ecological color variation：生態的色彩変異
ecological community：生態（学）的群集
ecological condition：生態的条件
ecological disturbance：生態系の攪乱、生態学的攪乱
ecological divergence：生態の多様化、生態（学）的分岐

ecological engineering：エコロジカルエンジニアリング、生態工学
ecological factor：生態の要因、生態の因子
ecological geography：生態地理学
ecological model：生態（学）モデル、生態系モデル
ecological niche：生態（学）的地位、エコロジカルニッチ
ecological region：生態学的地域
ecological requirement：生態の要求、生態学的要求事項
ecological resurgence：生態的誘導多発生
ecological significance：生態的意義
ecological space：生態的空間
ecological specialization：生態的特殊化
ecological trait：生態的特性
ecology：生態、生態学
economic balance of deciduous forest：落葉林の経済収支
economic injury level：経済的被害許容水準（EIL）
ecosystem：生態系、エコシステム
ecosystem engineer：生態系エンジニア
ecosystem service：生態系サービス
ecosystem-led approach：生態系主導のアプローチ
ecotype：生態型
ectopic eyespot：転位した眼状紋、異所的眼状紋
ectotherm：外温生物
ectothermic：変温の、外温の
edge effect：林縁効果、エッジ効果、周縁効果
edge of wing：翅端
editor：編集者
EDTA：エチレンジアミン四酢酸

(EthyleneDiamineTetraacetic Acid)
education：教育
effect of abundance：個体数効果
effect of acclimation period：馴化期間の効果
effect of continuous development：発育継続の効果
effect of inoculation：植氷効果、接種効果
effect of light interruption：光中断効果
effect of prior residence：先住効果
effect of spatial aggregation：空間的集団効果
effect of starvation：絶食の影響
effect of stem density：樹幹密度効果、幹数密度効果、本数密度効果
effective cumulative heat：有効積算温量
effective number of locus：遺伝子座の有効数
effective spectrum：有効スペクトル
effective temperature：有効温度
effective wavelength：有効波長
effector：効果器
efficiency of conversion of food into offspring：次世代への餌の転化効率
EGF：表皮成長因子、上皮細胞成長因子、上皮細胞増殖因子、上皮成長因子（Epidermal Growth Factor）
egg：卵（らん、たまご）
egg cannibalism：卵の共食い、卵食
egg cluster：卵塊
egg diapause：卵休眠
egg diploidization：卵の二倍体化
egg formation：卵形成
egg hatch：卵の孵化
egg hatching rate：卵孵化率

egg hatching success：卵の孵化成功率
egg mortality：卵死亡率、卵死亡数
egg overwintering：卵越冬
egg production：卵の生産
egg protein：卵タンパク質
egg stage：卵期
egg-diapausing species：卵休眠種
egg-laying：産卵
egg-load assessment：卵数の評価
egg-shell：卵殻（らんかく）
egg-to-adult developmental time：卵から成虫までの発育時間
EIA：環境影響評価（Environmental Impact Assessment）
EIL：経済的被害許容水準（Economic Injury Level）
ejaculate：射精する、射精、精液
ejaculatory duct：射精管
El Nino：エルニーニョ現象〔スペイン語〕
elaborate hypothesis：精巧な仮説
electroantennogram：触角電図、触角電図法
electromorph：電気泳動パターン、電気泳動型特異性
electrophoresis：電気泳動（法）
electrophysiological assay：電気生理学的方法、電気生理学的測定
electrophysiological response：電気生理学的反応
elementary color-coded neuron：色情報をコードする要素的なニューロン
elevated risk of population extinction：個体群絶滅の高まっているリスク
elevated threshold：上昇した閾（いき）値
elevation：高度、標高
elevational shift：高度変動、標高変動

elicit：解発する、誘起する、顕在化する
elicitor：エリシター
elide：脱字を施す、取り除く
elimination of anterior eyespot：前方眼状紋の切除
ellipsoidal eyespot：楕円体状の眼状紋
elongate：長く伸びる
elongate photoreceptor cell：細長い光受容細胞
elongation：伸長部、伸長
elongation factor：エロンゲーションファクター、伸長要素
elucidate：解明する
elucidation：解明
elusive quantity：捉えどころのない数量
elytron：翅鞘（ししょう）、さやばね
em-：中に入れる -、にする -〔ギリシャ語〕
Embioptera：シロアリモドキ目、紡脚目
embryogenesis：胚形成、胚発生
embryology：発生学
embryonic development：胚発育
embryonic mortality：胚死亡率、胎児死亡率
embryonic sex determination：胚の性決定
emendation：修正、修正名
emerge：羽化する
emerge from an egg：孵化する
emergence：羽化（うか）、孵化（ふか）、発生
emergence equalization：発生時期の斉一化
emergence rate：羽化率
emergence timing of cell row：細胞列の出現時期
emergency：創発、創発性
emergent：巨大木、抽水植物
emerging adult：羽化成虫

emerging form：新生の型
emerging species：新生種
emigrant：移出種
emigrate：移住する
emigration：移出、移住して行くこと、渡り、回遊
emigration rate：移出率
emperor butterfly：コムラサキ属の蝶
Emperors：コムラサキ亜科
empirical asymptote：実証的漸近
Empirical Bayes approach：経験ベイズ法
empirical evidence：実証の証拠
empirical research：実証研究
empirical result：実証的な結果、実験に基づいた結果
empirical test：実証試験
empodium：エムポディウム、爪間突起〔ポルトガル語〕
empty case：抜け殻
empty habitat：空き生息地
en：エングレイルド遺伝子（*engrailed* gene）
en-：中に入れる -、にする -〔ギリシャ語〕
end of light period：明期終了
end sequence：末端配列
endanger：衰亡の危機にさらす
endangered species：絶滅危惧種
endemic：固有の
endemic organism：固有生物
endemic species：固有種
endemic species to Japan：日本特産種、日本固有種
endemic to Japan：日本特産
ending：語尾
endo-：内 -〔ギリシャ語〕
endocrine factor：内分泌要因

endocrine mechanism：内分泌機構
endocrine organ：内分泌器官
endocrine regulation：内分泌調節
endocrine substance：内分泌物質
endocrine system：内分泌系
endocuticle：内クチクラ、内表皮(ないひょうひ)、内原表皮、エンドクチクラ
endogeneous circannual clock：内因的な概日時計、内因性リズム
endogenous octopamine receptor：内在性のオクトパミン受容体
endoparasitism：内部寄生
endophagous：内食性の
endosymbiont：細胞内共生細菌、細胞内共生微生物、内部共生微生物
endosymbiotic microorganism：細胞内共生微生物
endothermic：内温性の、温血の
enemy-free space：天敵不在空間
energy demand：エネルギー需要
enforcement：補強、強化
engrailed：エングレイルド遺伝子、波形縁の
enhancement：強化
enlarged region：拡大した領域
enlarged trachea：伸張した気管、拡張した気管
enlarged ventral eyespot：腹側の伸長した眼状紋列
enlargement：拡大図
enocytoid：エノシトイド(昆虫血球細胞の一種)
enrich：豊かにする
entail：課す
enthusiast：愛好家
entire experimental period：全実験期間

entomo-：昆虫-〔ギリシャ語〕
entomogenous：昆虫寄生性(の)
Entomological Society of Japan：日本昆虫学会
entomologist：昆虫学者
entomology：昆虫学
entomopathogenic：昆虫病原性(の)
entomophagous：食虫性(の)
entomophilous flower：虫媒花
entrainment：同調
entrenched：確立した
envelope：三角紙、エンベロープ、包み
environment：環境
environment cue：環境刺激
environmental change：環境変化、環境変動
environmental condition：環境条件
environmental conservation：環境保護、環境保全
environmental determinant：環境決定要因
environmental extreme：過酷な環境
environmental factor：環境要因
environmental fluctuation：環境変動
environmental gradient：環境傾度、環境勾配
environmental health：環境保健
environmental impact assessment：環境影響評価
environmental movement：環境保護運動
environmental period：環境周期
environmental protection：環境保全
environmental rearing condition：環境的飼育条件
environmental regulation：環境支配
environmental resistance：環境抵抗

environmental risk：環境リスク
environmental sensitivity：環境（的）感受性、環境的敏感性
environmental stochasticity：環境的確率性、環境変動の確率性
environmental variation：環境変動、環境多様性
environmentally conscious population：環境意識が高い国民
environmentally sound：環境への影響が少ない、環境にやさしい
enzymatic activation：酵素的活性化
enzyme：酵素
Eocene series：始新統
eon：累代、イーオン
Ephemeroptera：カゲロウ目、蜉蝣目
epi-：上 -〔ギリシャ語〕
epicuticle：上クチクラ、外表皮（がいひょうひ）、エピクチクラ
epidemic：伝染病、流行病
epidermal cell：表皮細胞、上皮細胞
epidermal cell layer：表皮細胞層
epidermal cell sheet：表皮細胞シート、表皮細胞層
epidermal growth factor：表皮成長因子、上皮細胞成長因子
epidermal organ：表皮器官
epidermal response：表皮反応
epidermis：表皮、真皮、表皮細胞
epigamic behavior：求愛行動
epigenetics：エピジェネティクス、後成学
epimicrospectrophotometry：エピ顕微分光測光法
epimorphic field：全射域
epiphysis：葉状片
epiproct：肛上板、肛上片

epistasis：エピスタシス、上位（性）、上下位性
epistatic incompatibility：上位不和合性
epistatic sex determinant locus：上位性決定遺伝子座
epistatic shutter allele：上位シャッター対立遺伝子
epizootic：流行性
epizootiology：流行病学（伝染病学）
epoch：世（地質時代の「せい」）
equality：斉一、均質
equalization：斉一化、均質化
equally：等しく
equator：赤道
equatorial：赤道直下の
equilibrium density：平衡密度
equipositional value line：等位置価で描かれた線
era：代（地質時代の「だい」）
eradicate：撲滅する
eradication：撲滅、根絶
eradication campaign：根絶キャンペーン
erase：消失する
erode：侵食する
erosion：低下
erratic suitability：変動性
error：過誤、ミス、誤り
eruciform：毛虫状の、イモ虫型
erythropterin：エリスロプテリン
escape behavior：逃避行動
escape mimicry：逃避擬態
escape response：回避反応
Escherichia coli：大腸菌
esophagus：食道
ESS：進化的安定戦略、進化的に安定な戦略（Evolutionarily Stable Strategy）

EST sequence：EST 配列（Expressed Sequence Tag sequence）
establish：設立する、確立する
established invasive alien species：定着してしまった侵略的外来種
establishment：定着、定着性、確立
establishment ability：定着力
esterifying acid：エステル化する酸
estimated distance：推定距離
estimated probability：推定確率
estimation of diversity index：多様性指標の推定、多様性指数の推定
estivation：夏眠、夏休眠〔英語〕
estuary：河口
ETH：脱皮刺激ホルモン（Ecdysis Triggering Hormone）
ethanol：エタノール
ethanol-precipitated DNA sample：エタノール沈殿させた DNA サンプル
ethical：倫理的
Ethiopia region：エチオピア区
ethyl acetate：酢酸エチル
etymology：語源学
eumelanin：真性メラニン、真正メラニン
eupyrene sperm：有核精子
eupyrene spermatozoon：有核精子
Eurasian continent：ユーラシア大陸
Eurasian origin：ユーラシア起源
eusociality：真社会性
evaginate：外転する、裏返しにする
evasive flight maneuver：逃避の飛翔行動作戦
evasive mimicry：逃避擬態
evasive response：逃避反応
evenness：均等度、均衡度
evenness of relative abundance distribution：相対的個体数分布の均等度
eventual result：終局の結果
evergreen broadleaf forest：常緑広葉樹林
evergreenness：常緑性
eversible tubercle：反転性伸縮突起
evo-devo：進化発生学、エボ - デボ
evolution：進化
evolution of butterfly foodplant：蝶の食性の進化
evolution of diversity：多様性の進化
evolution of migration rate：移動率の進化
evolutionarily stable strategy：進化的安定戦略
evolutionary arm race：進化的軍拡競争
evolutionary biology：進化生物学
evolutionary change：進化的変化
evolutionary consequence：進化的影響、進化的結果
evolutionary convergence：進化的収斂、進化的収束
evolutionary distance：進化距離
evolutionary inertium：進化的慣性
evolutionary optimum：進化的最適条件
evolutionary plasticity：進化的可塑性
evolutionary process：進化過程
evolutionary radiation：進化的放散
evolutionary relationship：進化的類縁関係、進化的関係
evolutionary rescue：進化的救済、進化的レスキュー
evolutionary response：進化的応答、進化的反応
evolutionary sequence：進化的配列
evolutionary significance：進化的意味、

進化的意義
evolutionary spiral：進化的スパイラル
evolutionary stable strategy：進化的安定戦略
evolutionary trajectory：進化軌跡、進化軌道
evolutionary trapping：進化上のトラップ、進化的トラップ、進化的罠
ex-：外へ-、無-、非-〔ラテン語〕
exacerbate：激化させる、悪化させる
exact probability：正確確率
examination of lectotype：タイプ標本（レクトタイプ）の調査
excess larval developmental time：幼虫の過剰な発育時間
exchange of information：情報交換
excluded：除外された
excluded name：除外名
excretion：排泄
excretory system：排泄系
exemplify：例証する
exhibit：示す、提示する
existence of territorialism：ナワバリ制の存在
existing population：既存の個体群
exo-：外-〔ギリシャ語〕
exocrine secretion：外分泌、外分泌物
exocrine substance：外分泌物質
exocuticle：外クチクラ、外原表皮、エキソクチクラ
exodus：移住
exogenous ammonia：外生アンモニア
exon：エクソン
exophagous：外的な送り装置
exoskeleton：外骨格
exotic：外来の

exotic garden plant：移入種の園芸植物
exotic species：外国種、外来種
expanding northern population：北方に拡大する個体群
expense of host fitness：宿主の適応度犠牲
experimental and correlative studies：実証的相関研究、実証的相関解析
experimental attention：実証的関心
experimental design：実験計画
experimental evidence：実験的証拠
experimental introduction：実験の導入
experimental period：実験期間
experimental population：実験個体群
experimental productivity gradient：実証的生産性勾配
experimental study：実験的研究、実験研究
exploitation：搾取作用
exploitative competition：消費型競争
explosive adaptive radiation：急激な適応放散
exposure：曝（さら）すこと、晒すこと
expression：発現、表現
expression level：発現量、発現レベル
expression of female phenotype：雌の表現型発現
expression of seasonal form：季節型発現
expression pattern：発現パターン
expression plasticity：表現可塑性
expression regulation：発現調節
expression signature tag：EST 解析
expression tendency：発現傾向
expressional regulation：発現調節
extant：現存している、現生、現存
extant population：現存個体群
extant species：現生種
extant variety：在来種

extend laterally：外側に伸長する
extended point：突出部
extensive survey：広範囲な調査、大規模な調査
extent of wing melanization：翅の黒化の程度
external：外(がい)、外部の
external anatomy：外部組織、外部の解剖学的構造、外部構造
external body feature：外部構造
external character：外部形質
external coincidence model：外的符合モデル
external genitalia：外部生殖器
external morphology：外部形態(学)
external stimulus：外部刺激、外界の刺激
extinct：絶滅した、絶滅
extinct species：絶滅種
extinction：死滅、絶滅
extinction and colonisation dynamics：絶滅 - 定着動態、絶滅 - 移入動態
extinction rate：絶滅率
extinction risk：絶滅リスク
extirpated：根絶した
extra molt：過剰脱皮
extra-：領域外の -、範囲外の -〔ラテン語〕
extracellular recording：細胞外記録
extracted DNA：抽出 DNA
extrafloral nectary：花外蜜腺
extraocular photoreceptor：視覚外光受容器
extraordinary number：非常に大きな個数、途方もない個数
extraordinary proportion：異常な割合、非常に大きな割合
extrapolation：外挿
extreme condition：極限状態
extreme unevenness：極端な不均等、極端な不均一
extrinsic：外因的、(環境的)
exuded tree sap：滲み出た樹液
exuvium：脱皮殻(だっぴから)
eye：眼
eye cap：目蓋
eye-antennal disc：眼 - 触角原基
eyeshine：暗視眼(動物の目が暗闇の中で光る効果)
eyespot：眼状紋(がんじょうもん)、目玉模様、眼状斑点
eyespot absent：眼状紋の消失
eyespot focus：眼状紋フォーカス
eyespot formation：眼状紋形成
eyespot size：眼状紋サイズ
eyespot-associated gene：眼状紋関連遺伝子

f

F：前翅(ぜんし)(Forewing)
f.：品種(form/forma)、「『属名 種名 f. 品種名』の形式で表記」
FA：変動非対称性(Fluctuating Asymmetry)
fabric of nature：自然の構造
face：顔面
facet：個眼(面)
factorial design：要因配置計画
facultative：外因性、条件的、機会的、任意の
facultative association：任意共生的関係
facultative diapause：外因性休眠
facultative migration：随意移動
facultative mutualism：任意的相利共生、条件的相利共生、不偏性相利共生

facultative parthenogenesis：機会的単為生殖
fall form：秋型
fall in：分類される
fallen fruit：落果
falling together：共倒れ
fallow：休耕地
false alpine butterfly：偽の高山蝶
false eye：擬眼、擬目
false head：擬頭
false negative：偽陰性
falsification：反証
fam.：科（family）
fam. nov.：新科名（family novus）
family：科（分類階級の「か」）
family group：科階級群
family name：科名
family--order ratio：科数 - 目数の比
family-group name：科階級群名
far field pressure component of sound：遠距離場の音の圧力成分
Faraday cage：ファラデーケージ、ファラデー箱
fare：暮らす
farming：耕作化、養殖
farmland：農地、農村地帯、農村地域
farmland habitat：農地性生息場所
fascinating：興味をそそられる
fast-changing environment：急速に変化する環境
fast-flying：すばやく飛ぶ
fat：太らす
fat body：脂肪体、肥満体
fatally damage：致命的に損傷する
fauna：動物相、ファウナ
faunal region：動物地理区

favea：窩、中心窩
favor：好む
favorable breeding condition：有利な繁殖条件、良好な繁殖条件
favorable host plant：好都合な寄主植物
feather-like marking：羽毛状の斑紋
feathered spine：羽毛のようなトゲ
feature：特徴
fecundity：蔵卵数、産卵数、繁殖力、産卵力、生殖力、妊性、稔性
feed：食事をする
feed conversion efficiency：餌の転換効率
feed on：食する
feeding：吸蜜中、摂食中
feeding activity：摂食活性
feeding attractant：摂食誘引物質
feeding behavior：摂食行動、摂食活動
feeding cut off：摂食停止
feeding deterrent：摂食阻害物質
feeding differentiation：食性分化
feeding period：摂食期
feeding site：摂食場所、餌場
feeding stimulation：摂食刺激
female：雌（めす）、♀、メス
female butterfly：母蝶
female defense polygyny：雌防衛型の一夫多妻
female eclosion：雌の羽化
female embryo：雌（の）胚
female genitalia：雌性生殖器
female heterogametic：雌ヘテロ型、雌異型配偶子型
female income：雌の収入
female model：雌モデル
female monogamy：雌性単婚性
female parent：母親

female phenotype：雌の表現型
female progeny：雌の子孫
female-heterogametic chromosomal constitution：雌ヘテロ型性染色体構成
female-heterogametic sex chromosomal system：雌ヘテロ型性染色体システム、雌ヘテロ型性染色体様式
female-informative：雌に有益な
female-specific molecular mechanism：雌に特有の分子機構
female-specific organ：雌特有の器官
feminization：雌性化、雌化
feminize：雌化する
femur：（複．-ra）、腿節（たいせつ）〔ラテン語〕
fermenting：発酵した
fermenting fruit：発酵した果実
fertile：繁殖力のある、妊性、稔性の
fertile egg：有精卵、受精卵
fertilise：肥沃にする、受精させる
fertility：妊性、稔性、生殖能、受精率
fertilization：受精
fertilization rate：受精率
fertilize：受精させる
Feulgen's reaction：フォイルゲン反応
FH：両翅（Forewing and Hindwing）
fibular：腓（ひ）側、腓骨の
FID：水素炎イオン化型検出器（Flame Ionization Detector）
field：野外、フィールド、圃場、分野
field assay：野外検定
field assistance：フィールド支援、野外支援
field experiment：野外実験
field memory recorder：フィールドメモリーレコーダー
field study：フィールド研究、野外研究
field work：野外観察
field-caught female：野外採集した雌
filament：触角、触手、糸、花糸、フィラメント
filamentous：繊維質の
filial：第〇世代の、交配世代の
filter paper：濾紙（ろし）
finch：フィンチ（スズメ科の小鳥）
fine：罰金
fine forceps：（ハサミムシなどの）微細なハサミ、微細な鉗子
fine structure：微細構造
fine-scale genetic mapping：詳細な遺伝子地図
fine-scale mapping：詳細なマッピング、細密なマッピング
fine-scale spatial mosaic：微細スケールの空間（的）モザイク
first author：筆頭著者、第一著者
first comer：最も早く飛来した個体
first filial generation：雑種第一代目
first generation：第一世代
first generation adult：第一世代成虫
first instar：初齢、一齢
first line of defence：防衛の最前線
first reviser：最初の改訂者、第一校訂者
first segment：第一節
first-generation hybrid：第一代雑種、第一世代の雑種
Fisher's exact test：フィッシャーの正確確率検定
fishing line：釣糸
fitness：適応度、フィットネス
fitness parameter：適応度因子、適応度パラメータ、適応パラメータ
fixation：固定

fixation by elimination：消去法による固定
fixation of mimicry gene：擬態遺伝子の固定
fixative：固定剤、定着剤、色留め剤
fixed day：定日型（ていじつがた）
fixed divergent：固定、固定分岐
fixed nonsynonymous substitution：非同義性固定置換
fixed synonymous substitution：同義性固定置換
fixed time diapause：定時的休眠、定時休眠
flagellum：鞭節（べんせつ）〔ラテン語〕
flagship insect：象徴昆虫
flagship species：象徴種
flame ionization detector：水素炎イオン化型検出器
flank：側面を接する
flapping wing：バタバタさせている翅、羽ばたきをする
flash coloration：閃光色
flat cover：扁平な袋
flattened club：平べったい先端
flattened spheroidal surface of wing：扁平な楕円体状の翅表面
flavone：フラボン
flavonoid：フラボノイド
flavonoids：フラボノイド系色素
flavonol：フラボノール
flavonol glycoside：フラボノールグリコシド、フラボノール配糖体
fleeing point：逃避点
fleshy：肉質の
fleshy spine：肉質の突起
flier form：移動型
flight：飛翔、移動飛翔

flight ability：飛翔能力、飛行能力
flight activity：飛翔活動
flight behavior polymorphism：飛翔行動多型
flight boundary layer：飛行境界層
flight capacity：飛翔能力
flight distance：移動距離
flight morphology：飛翔関連形態
flight muscle：飛翔筋
flight muscle mass：飛翔筋重量
flight muscle polymorphism：飛翔筋多型
flight path：移動経路
flight period：発生期間
flight polymorphism：移動多型
flight sound：飛翔音
flight sound of avian predator：鳥捕食者の飛翔音
flight space：飛翔空間
flight speed：飛翔速度
flight time：飛翔時間
flight-related behavior：飛翔関連行動
flood mitigation：洪水緩和
flora：植物相、フローラ
floral belt：植生帯、植物帯
floral nectar：花蜜
floral organ formation：花器形成
floral scent：花の香気成分、花の発散香気成分
floral volatile：花の揮発性成分、花の揮発性物質
floristically：植物相的に
flour beetle：小麦粉につく甲虫、コクヌストモドキ
flourish：繁茂する
flower constancy：定花性、花選好性、花種選好性

flower garden：花壇、花園
flower-like spider：ハナグモ
flower-visiting behavior：訪花行動
flowering individual：顕花個体
flowering plant：顕花植物
fluctuate：変動する
fluctuating asymmetry：変動非対称性、（左右）対称性のゆらぎ、ゆらぎの非対称性
fluctuating temperature treatment：変温処理
fluctuation：変動、ゆらぎ
fluid：流動体、液体
fluid secretion：液体分泌、水分分泌
fluorescent bulb：蛍光灯電球
fluorescent paint：蛍光塗料
fluorescent-labelled primer：蛍光標識したプライマー
flutter：羽ばたきをする、飛翔する
flutter response：はばたき反応
flutter wing：翅をパタパタと開閉する
fluttering speed：飛翔速度
fly：飛ぶ、ハエ
fly vertically from low to high altitude：平地と高標高地間を垂直に移動する
fly way：蝶道、飛行経路
flycatcher：ヒタキ（小鳥の一種）
flying type：飛翔型
focal ablation：フォーカスの切除
focal graft：フォーカスの移植
focal signal：フォーカスシグナル、焦点信号
focus：フォーカス、焦点
foe：敵、かたき、敵対者
fold：折りたたむ、ひだ、褶
folded：よじれた

foliage：群葉、葉、葉物
foliage type for oviposition：産卵に適した葉の種類
folivore：葉食動物
follicle：濾胞
follicular cell：濾胞細胞
following day：次の日、翌日
food abundance：食物の豊富さ
food chain：食物連鎖、食物網
food gathering：採食
food habit：食性
food plant：食草、食餌植物
food recruitment：食物調達、食物確保
food shortage：食物不足
food supply：食物供給
food web：食物連鎖、食物網
foodplant change：食草転換
fool：騙す
forage：食糧をあさる、食糧を入手する
foraging：採餌
foraging behavior：採餌行動
foraging preference：採餌選好性
foraging site：食料を入手する場所
foraging strategy：採餌戦略
fore leg：前脚(ぜんきゃく)、前肢(ぜんし)
fore-：前 -〔ラテン語〕
forefront：最前線
foregut：前腸
foreign pest：外来性害虫
foreleg：前脚(ぜんきゃく)、前肢(ぜんし)
foreleg folded on its thorax：胸部に折り畳まれている前脚
foreleg magnified：前脚の拡大
forest：森林
forest and relictual genus：森林遺存属

forest and wide-distributed butterfly：森林性広域分布蝶
forest and wide-distributed genus of Southeast Asian origin：東南アジア森林広域分布属
forest butterfly：森林性蝶
forest canopy habit：林冠性
forest clearing：森林開拓地
forest edge：林縁部、森林の周縁域
forest environment：森林的環境
forest fire：森林火災
forest floor：林床
forest floor habit：林床性
forest gap：森林ギャップ
forest margin：林縁
forest margin area：森林の周縁域
forest matrix：森林マトリックス
forest non-gap：林冠
forest species：森林性種
forest, savanna and wide-distributed genus：森林サバンナ広域分布属
forest-dwelling species：森林性種
forestry：林業
foretarsal receptor：前脚跗節感覚器、前脚跗節受容器
foretarsus：前脚跗節
forewing (HW)：前翅（ぜんし）、上翅（じょうし）
forewing band：前翅の帯
forewing bud of pupa：蛹の前翅芽、蛹期の前翅原基
forewing color：前翅の色
forewing color gene：前翅の色彩遺伝子
forewing fringe color：前翅縁毛色
forewing length：前翅長
forewing overlying hindwing：後翅を覆っている前翅

forewing patch：前翅の大斑点
forewing structure：前翅の構造
forewing upperside：前翅表面
forewing's apex：前翅の最上端部
form：型、形
form ectopically：正常でない位置に起こる形
formation mechanism：形成機構
formation of ectopic eyespot：転位した眼状紋の形成
formation of garter：帯糸形成
formation process：形成過程
formerly-declining butterfly：以前に衰亡しつつあった蝶
forward and reverse primers：フォワードプライマーとリバースプライマー、順方向プライマー及び逆方向プライマー
fossil：化石
founder：創始者
founder effect：創始者効果
founder event：創始者事象
founder selection：創始者選択
fragmented landscape：分断された地形、分断化された景観
fragmented population：分断化された個体群、断片化された個体群
fragrance composition：芳香組成物
frass：糞粒、糞
free glycine：遊離グリシン、遊離型グリシン
free progress：自由進行
freeze intolerance：非耐凍性
freeze susceptible：非耐凍性、凍結感受性
freeze tolerance：耐凍性
freeze-intolerant insect：非耐凍型昆虫、

非耐凍性昆虫

freeze-intolerant species：非凍型種、非耐凍性種

freeze-tolerant species：耐凍性種

freezing injury：凍傷

frenulum：翅鉤(はねかぎ)、翅刺(しきょく)、翅棘(しきょく)

frequency：周波数、振動数、頻度

frequency discrimination：周波数弁別

frequency distribution：頻度分布

frequency of adult：成虫の頻度

frequency of the yellow allele：黄色対立遺伝子頻度

frequency-dependent competition：頻度依存的な競争

frequency-dependent predation：頻度依存的な捕食、頻度依存性捕食

frequency-dependent selection：頻度依存選択

fresh nectar：新鮮な果汁

fresh weight：生体重

freshly emerged specimen：生き生きと出現した標本

fright tactic：脅かし戦術

fringe：縁毛、縁毛帯、フリンジ

fringe color：縁毛色

Fritillary：ヒョウモン

fritillary marking of black spot on an orange ground：オレンジ色の地色に黒斑のあるヒョウモンチョウ特有の翅模様

frog：カエル

frons：(複．frontes)、額、ひたい、額板、前頭〔ラテン語〕

front pair of leg：前脚、前肢

frost：氷点下、氷結

fruit fly：ミバエ、ショウジョウバエ

fruit mimic：果実に擬態

fruit orchard：果樹園

fruit-feeder：果物食者

full brother：同父母の兄(弟)、実の兄(弟)

full protection：完全保護

full sibling：全部の近縁種、全同胞種

full-length：全長

full-length sequence：全長配列

fully grown：終齢

fully protected species：完全保護種、完全に保護されている種

fully-formed larva：成体が完成した幼虫

fumigant：燻蒸剤(くんじょうざい)

functional divergence：機能的な分岐、機能分化、機能多様性、機能発散

functional female：機能的な雌、機能性雌

functional insect ear：昆虫の機能的な耳

functional interrelationship：機能分担

functional organization：機能的機構、機能的構成、機能的構築

functional response：機能の反応、機能的反応

functionally related gene：機能的に近縁な遺伝子

fundamental niche：基本ニッチ、基本生息場所

fundatrix：幹母

fungus：(複．-gi)、糸状菌、カビ

funicle：繋節

fuse：融合する

fused eyespot doublet：融合して一つになった眼状紋の対

fused microvillar membrane：融合した微絨毛膜

fusion of gradients：勾配群の融合

g

G protein-coupled receptor：G タンパク質結合受容体、G タンパク質共役型受容体
G-protein：G タンパク質
g.：属（genus）
g. nov.：新属名（genus novus）
G6PDH：グルコース-6-リン酸デヒドロゲナーゼ（Glucose-6-Phosphate DeHydrogenase）
gain of function：機能亢進、機能獲得
gallery forest：ガレリア森林（サバンナなどの川沿いの帯状林）、拠水林
gamete：配偶子
gamete duplication：生殖核倍加型
gametic isolation：配偶子隔離、配偶子単離
gamma distribution：ガンマ分布
gamma irradiation：γ線照射、ガンマ線照射
gamma-distributed rate：ガンマ分布比
ganglion thoracicum primum：第一胸節神経球
gap habitat：ギャップ生息地
garden tiger moth：ヒトリガ
garden warbler：ニワムシクイ
gas chromatography：ガスクロマトグラフィー
gas exchange：ガス交換
gate：ゲート
GC：ガスクロマトグラフィー（Gas Chromatography）
GC-MS：ガスクロマトグラフィー-マススペクトロメトリー（Gas Chromatography Mass Spectrometry）
gel：ゲル〔ドイツ語〕、ジェル〔英語〕
gen-：遺伝の-、遺伝子の-〔ギリシャ語〕
gen.：属（genus）
gen. nov.：新属名（genus novus）
gena：頬〔ラテン語〕
GenBank accession no.：遺伝子銀行受入れ番号、ジーンバンクアクセス番号
gender：性別、性
gender agreement：性の一致
gender ending：性語尾
genders：雄雌、雄と雌、ジェンダー
gene：遺伝子
gene Cubitus-interruptus：肘脈中断遺伝子（翅脈が途切れている）
gene duplication：遺伝子重複
gene expression：遺伝子発現
gene expression analysis：遺伝子発現解析
gene expression cascade：遺伝子発現カスケード
gene expression study：遺伝子発現研究
gene flow：遺伝子流動、遺伝子の流れ、遺伝子交流、遺伝的交流
gene genealogy：遺伝子系図、遺伝子の系譜、遺伝子系統学、遺伝子系図学
gene knockout：遺伝子欠損、遺伝子破壊
gene mutation：遺伝子突然変異
gene order：遺伝子配列順、遺伝子順序
gene product：遺伝子産物
gene region：遺伝子領域
gene sequence：遺伝子配列
gene *wingless*：無翅遺伝子
gene-specific reverse primer：遺伝子特異的逆転写用プライマー
genealogical change：系統変化
genealogical tree：系統樹

genealogy：系統、系譜
general epidermal cell：一般的な表皮細胞
generalist：ジェネラリスト、万能家
generalist feeding：雑食性
generalist species：普遍性の種
generalization：一般化、普遍化
generalized linear model：一般化線形モデル
generation：世代
generic name：属名
generic richness：属数
genesis：発生
genetic analysis：遺伝子解析、遺伝子分析
genetic and biochemical approach：遺伝生化学的手法
genetic architecture：遺伝的構成
genetic association：遺伝相関
genetic background：遺伝的基盤、遺伝的背景
genetic basis：遺伝的基盤、遺伝的基礎
genetic cause：遺伝的原因
genetic character of metapopulation：メタ個体群の遺伝の形質
genetic control：遺伝的制御、遺伝的防除
genetic control of plasticity：可塑性の遺伝的制御
genetic correlation：遺伝相関
genetic coupling：遺伝的結合、遺伝的連関
genetic covariance：遺伝共分散
genetic diapause：遺伝的休眠
genetic difference：遺伝的差異
genetic differentiation：遺伝的分化
genetic distance：遺伝(的)距離、地図距離
genetic distinction：遺伝的相違、遺伝的差異、遺伝的区別
genetic distinctiveness：遺伝的特殊性
genetic divergence：遺伝的分化
genetic diversity：遺伝的多様性
genetic drift：遺伝的浮動
genetic effect：遺伝の影響
genetic factor：遺伝的要因
genetic homology：遺伝的相同性
genetic incompatibility：遺伝的不一致、遺伝的不和合性
genetic integrity：遺伝的完全性
genetic level：遺伝的レベル
genetic linkage map：遺伝子連鎖地図、遺伝の連鎖地図
genetic linkage of sexual isolating trait：性的隔離に関する形質の遺伝の連鎖
genetic load：遺伝的負荷、遺伝荷重、遺伝的加重
genetic male：遺伝的雄
genetic mapping：遺伝子マッピング、遺伝子地図作製
genetic marker：遺伝マーカー、遺伝子マーカー、遺伝標識
genetic mechanism：遺伝機構
genetic model：遺伝モデル
genetic mosaicism：遺伝的モザイク現象
genetic polyphenism：遺伝的多型
genetic relationship：遺伝的交流、遺伝的関係
genetic speciation：遺伝的種分化
genetic statistics：遺伝統計学
genetic stochasticity：遺伝的確率性
genetic structure：遺伝的構造
genetic task specialization：遺伝的な役割分業
genetic variability：遺伝的変異性、遺伝変異性

genetic variance：遺伝分散
genetic variant：遺伝的変異体
genetic variation：遺伝的変異、遺伝性変異、遺伝的の変動
genetically associated：遺伝的に相関した
genetically determine：遺伝的に決定する
genetically determined preference：遺伝的に決まっている選好性
genetically determined trait：遺伝的に決まっている形質
genetically female part：遺伝的雌部位
genetically male individual：遺伝的雄個体群
genetically male part：遺伝的雄部位
genetically-diverse population：遺伝的に多様な個体群
genetics：遺伝学、遺伝的特徴
genital papilla：側唇、生殖瘤状突起
genital part：生殖部位
genital photoreceptor：尾端光受容器
genitalia：ゲニタリア、生殖器、交尾器〔ラテン語〕
genitalia vial：ゲニタリアチューブ
genitive ending：属格語尾
geno-：遺伝の -、遺伝子の -〔ギリシャ語〕
genome：ゲノム
genome average relationship：ゲノム平均の関係
genome coverage：ゲノムのカバー率
genome scale：ゲノム尺度
genome sequencing：ゲノム配列
genome-wide：ゲノム規模の
genome-wide introgression：ゲノムワイドの遺伝子移入
genomic block：ゲノムブロック
genomic imprinting：ゲノムインプリンティング、ゲノム刷り込み
genomic incompatibility：遺伝的不和合性、遺伝的不一致
genomic location：ゲノム位置、遺伝子位置
genomic region：ゲノム領域
genomic resequencing：ゲノム塩基配列の再解析
genomic resource：ゲノムリソース、ゲノム資源
genomic scaffold：ゲノムスキャフォールド、ゲノム骨組
genomic study：ゲノム研究
genomic tool：ゲノミクスの手法
genotype：遺伝子型、遺伝型、遺伝子型を決定する
genotype-environment interaction：遺伝子型 - 環境相互作用、遺伝子型と環境との相互作用
genotyping：遺伝子型同定、遺伝子型解析、遺伝子型判定
Genoveva azure butterfly：オオヤドリギシジミ
genus：（複 . genera)、属（分類階級の「ぞく」)〔ラテン語〕
genus group：属階級群
genus name：属名
genus of neotropical butterfly：新熱帯区に生息する蝶の属
genus-group name：属階級群名
geographic distance：地理的距離
geographic distribution：地理的分布
geographic isolation：地理的隔離
geographic mosaic：地理的モザイク
geographic population：地理的個体群
geographic radiation：地理的放散

geographic restriction：地理的制限、地理的制約
geographic speciation：地理的種分化、地理的な分化、地理的分化
geographic species：地理的種
geographic variation：地理的変異、地域変異
geographical background：地理的背景、地誌的背景
geographical cline：地理的クライン、地理的勾配
geographical distribution type：地理的分布型
geographical gradient：地理的勾配
geographical race：地理的系統、地理的品種、地理的亜種
geographical region：地理区
geographical resolution：地理的解像度
geographical space：地理的空間
geographically close：地理的に近傍な
geohistorical background：地史的背景
geological and historical background：地史的背景
geological succession：地史的遷移
geological time：地質時代
geological variation：地域変異、地史的変異
geophagy：土食、土壌食性
germ band：胚帯
germ cell：胚細胞、生殖系列細胞、生殖細胞
germinate：芽を出す
germination success：発芽成功
ghost authorship：ゴースト著者
ghost moth：コウモリガ
giant glial cell：巨大グリア細胞

giant redeye butterfly：コウモリセセリ
giant swallowtail butterfly：クレスフォンテスタスキアゲハ、オオタスキアゲハ
Giant-Skippers：イトランセセリ亜科
gift authorship：ギフトオーサーシップ
girdle：帯、帯糸(たいし)
girdle for support：支持糸
girdling：帯糸をかける
glacial epoch：氷期
glacial period：氷河時代、氷河期
glass knife：ガラスナイフ
glass microscope slide：顕微鏡用のスライドガラス
glassine envelope：グラシン（紙）の三角紙
gleaning bat：ウサギコウモリ
glial cell：グリア細胞
glial cell layer：グリア細胞層
global biodiversity loss：世界の生物多様性喪失
global environmental problem：地球規模の環境問題
Global Register of Invasive Species：グローバル侵入種登録簿
global warming：地球温暖化
globalization：グローバリゼーション、世界の規模
globe：地球
glucose：グルコース
glucose-6-phosphate dehydrogenase：グルコース-6-リン酸デヒドロゲナーゼ
glucosinolate：グルコシノレート、カラシ油配糖体
glutathione-S-transferase：グルタチオン・S・トランスフェラーゼ、グルタチオンS-転移酵素

glycerin：グリセリン
glycerol accumulation：グリセロールの蓄積
glycine-rich：高グリシン含有
glycogen：グリコーゲン
glycogen content：グリコーゲン含量
glycolysis：解糖系
gnathos (gn)：顎(がく)、あご
goblet cell：杯状細胞
gold annulus：金環
gold ring：金環
gold-drop helicopis butterfly：ミツオシジミタテハ
golden birdwing butterfly：キシタアゲハ
golden piper butterfly：チャオビタテハ
goldpalladium：金パラジウム、金パラ
golgi body：ゴルジ体
Goliath birdwing butterfly：ゴライアストリバネアゲハ
gonadal development：生殖腺の発育
Gossamer wings：シジミチョウ科
Gossamer-wing Butterflies：シジミチョウ科
GPCR：Gタンパク質結合受容体（G Protein-Coupled Receptor）
Gr：味覚受容体（Gustatory receptor）
Gr5a：味覚受容体遺伝子の一つ（Gustatory receptor 5a）
graceful appearance：美麗種
gradient：勾配、グラディエント、段階的変化
gradient profile：勾配プロフィール、勾配プロファイル
gradient variation：勾配変異
gradual：ゆるやかな
gradual change：漸進的変化

gradual evolution：漸進的進化
gradual tightening of linkage：連鎖を徐々に密にして行く
graft：移植組織、移植片、接ぎ木
grafted focus：移植したフォーカス
grafting experiment：移植実験
granulocyte：顆粒細胞(かりゅうさいぼう)
granulosis：顆粒病（昆虫ウイルス病の一種）
granulosis virus：顆粒病ウイルス
Grass Skippers：セセリチョウ亜科
grasshopper：キリギリス
grassland：草原地帯、草原、草地、牧草地
grassland and relictual genus：草原遺存属
grassland and wide-distributed genus：草原広域分布属
grassland butterfly：草原性蝶
grassland environment：草原的環境
grassland genus of American origin：アメリカ起源草原属
grassland genus of Eurasian origin：ユーラシア起源草原属
grassland habitat：草原性生息地
gravid：抱卵、受胎(じゅたい)
gravity：重さ、重量
grazing：放牧
greasy：油紙のような
great purple hairstreak butterfly：アメリカヤドリギシジミ（ヤドリギは幼虫の食餌植物）
Greater Sunda Islands：大スンダ列島
greatest common divisor：最大公約数
Greek：ギリシャ語
green beard effect：緑髭効果
green light：緑色光
green revolution：緑の革命

green-blue marking：緑青色斑紋
greenhouse：温室
greenhouse effect：温室効果
greenhouse gas：温室効果ガス
gregarious phase：群生相
gregarious roosting：集団帰塒(しゅうだんきじ)
gregariously：集合性があり、群れて
gregariousness：集合性、群居性、群生、群居
grid squares, 10km：10km グリッド四方
GRIs：グローバル侵入種登録簿(Global Register of Invasive species)
groove：溝
ground：地面
ground color：地色
ground scale：グランドスケール
group：集団、個体群、群、階級群
group selection：群淘汰、集団選択、群選択
grow：成長
growing period：生育期、成長期
growth：成長、成長量、成長率、生長
growth chamber：培養室
growth condition：成長条件、生育条件
growth day：発育日数
growth inhibitor：成長・発育を阻害する物質
grub：幼虫
Grylloblattodea：ガロアムシ目、欠翅目
GST：グルタチオン・S・トランスフェラーゼ(Glutathione-S-Transferase)
guava skipper butterfly：シロベリセセリ
guest authorship：ゲストオーサーシップ
guide mark：花標
Guiding Principles for the Prevention, Introduction and Mitigation of Impacts of Alien Species：外来種の影響の予防、導入、影響緩和のための指針原則
guise：見せかけ
gulching：待ち伏せ
gustatory neuron：味覚ニューロン、味覚神経細胞
gustatory organ：味覚器官
gustatory reception capacity：味覚受容能力
gustatory receptor：味覚受容体
gustatory sense：味覚
gustatory sensory hair：味覚感覚毛
gustatory stimulus：味刺激
gut purge：ガットパージ、脱糞、液状糞
gymnosperm：裸子植物
gynander：雌雄型
gynandromorph：雌雄型、ジナンドロモルフ、雌雄モザイク
gynandromorphism：雌雄モザイク現象
gypsy moth：マイマイガ

h

H：後翅(こうし)(Hindwing)
h after, 24：〜後24時間、24時間後
h old, 24：24時間まで、〜後24時間
habit：習性
habitant：生息生物
habitat：生息地、生息環境、生育地、自生地、ハビタット
habitat change：生息地変化
habitat destruction：生息地破壊
habitat factor：生息環境要因
habitat fragmentation：生息地の分断化
habitat islands：島嶼生息地
habitat isolation：生息地隔離

habitat loss：生息地の減少、生息地の消失、生息地の喪失
habitat management：生息地管理
habitat manipulation：生息場所の操作
habitat modification：生息地の改変、生息地の造成
habitat patch：生息地パッチ
habitat range：棲息圏、生息圏
habitat remnant：生息地の名残
habitat segmentation：生息地の分断化
habitat segregation：すみわけ
habitat specialist：生息場所特定者
habitat specialist butterfly：生息場所特定性の蝶
habitat specialist species：生息場所特定性の種
habitat specificity：生息地の特異性
habitat structure：生息地の構造、生息地の分布構造
habitat suitability：生息場所適性、生息地適正、生息環境適正、生息地適合性
habitat templet hypothesis：生息場所鋳型説
habitat-patch occupancy：生息地パッチの占有率
haemocyte：血球、血球細胞
haemoglobin：ヘモグロビン
haemolymph：血リンパ
haemolymph sample：血リンパのサンプル
hair：毛
hair tuft：毛束
hair-like：毛状の
hairpencil：ヘアペンシル、毛束
hairpencil dihydropyrrolizine：ヘアペンシルから分泌されるジヒドロピロリジン化合物

Hairstreaks：カラスシジミ亜科
Haldane centiMorgan：ホールデンセンチモルガン
Haldane's rule：ホールデインの法則、ホールデンの法則
half-life：半減期
hallucinogenic：幻覚作用
Hamilton rule：ハミルトン則、ハミルトンの法則
Hamilton's rule：ハミルトン則、ハミルトンの法則
Hampson's classification：ハンプソン式、ハンプソンの分類
hand pairing method：ハンドペアリング法
hand-pairing：ハンドペアリング
handling：取り扱い
handling time：処理時間
hapantotype：ハパントタイプ（標本）
haplodiploid sex determination system：半倍数性の性決定システム、単数二倍体の性決定様式
haplodiploidy：半倍数性
haplogroup：ハプログループ
haploid genotype：半数体の遺伝子型
haplotype：ハプロタイプ（"haploid genotype"の略語）、単相の
haplotype analysis：ハプロタイプ解析、分子系統解析
haplotype diversity：ハプロタイプ多様性
haplotype network：ハプロタイプネットワーク
harassment：干渉、ハラスメント
harassment activity：干渉行動
harbor：宿る、住みかとなる、ひそむ
hard leaf：硬い葉
hard rain：大雨

Hardy-Weinberg equilibrium：ハーディワインベルグ平衡
Hardy-Weinberg expectation：ハーディー‐ワインベルグ期待値
harmless species：無害種
harmony life：調和的な生活
harpe：ハルペ、側鉤器
Harvesters：カニアシシジミ亜科
hatch：孵化(ふか)する
hatched egg：孵化した卵
hatching success：孵化成功
haustellum：口吻(こうふん)、吸収管、吸管
hawk moth：スズメガ
head：頭部(とうぶ)
head shell：頭殻
head truncation：頭部切断
head width：頭幅
headspace：ヘッドスペース法
healthiest population：最も健康な個体群
hearing：聴覚
hearing function：聴覚機能
hearing organ：聴覚器官
heart：心臓(しんぞう)、背脈管
heat absorption：熱吸収
heat gain：熱取得効率
heat paralysis：熱麻痺(ねつまひ)
heated chain：高温系列
heathland：ヒース地帯、ヒースランド、荒地
heathland vegetation：ヒース地帯の植生
heavily-wooded landscape：うっそうと茂った森の景観、深い森の景観
Hebe：ゴマノハグサ
hectographing：ゼラチン版印刷
hedgehog：ヘッジホッグ(遺伝子発現)、ハリネズミ状紋

hedgehog signalling：ヘッジホッグシグナル伝達
hedgerow：垣根、生け垣
hedgerow management：低木の列管理、潅木管理、生け垣管理
hedylid moth：シャクガモドキ科の蛾
heliconiine butterfly：ドクチョウ属の蝶
Heliconius butterfly：ドクチョウ、有毒蝶
Heliconius cydno：シロオビドクチョウ
hemi-：半‐〔ギリシャ語〕
Hemimetabola：不完全変態する昆虫類
hemimetabolism：不完全変態
hemimetaboly：不完全変態
Hemiptera：カメムシ目、半翅目
hemivoltine：半化性(生活史が二年)、二年生の化性
hemo-：血‐〔ギリシャ語〕
hemocoel：血体腔
hemocyte：血球、血液細胞
hemolin：ヘモリン
hemolymph：血リンパ、血液
hemolymph ecdysteroid titer：血液中のエクジステロイド量、血リンパ‐エクジステロイド価
herb：ハーブ、草本
herbaceous plant：草本植物
herbicide：除草剤
herbivore：草食(性)動物
herbivore-induced plant volatile：植食者が誘導する植物の揮発成分、植食者誘導性植物揮発性物質
herbivorous lepidopteran：植食性チョウ目昆虫
herbivory：植食性、草食、植食
hermaphrodite：雌雄同体、雌雄同株、両性動物

hetero male：雄ヘテロ型

heterodimeric receptor：ヘテロ二量体を形成する受容体、ヘテロ二量体受容体、ヘテロダイマー受容体

heterodynamic：異動態的、季節別繁殖動態的、ヘテロダイナミック、（[注] 休眠ありの生活環）

heterodynamic type：異動態的な発生型、周期型

heterogeneity：異種混交性、異質性

heterogeneity of butterfly eye：蝶の眼の異質性、蝶の眼の多様性

heterogeneous environment：異質な環境

heterogeneous habitat：異質な生息地

heterogeneous selection：異種選択、異質選択、不均質選択

heterogeneous species：異質な種

heterogeneously expressed filtering pigment：不均一に発現したフィルタリング色素

heterospecific female：異種の雌

heterostyly：異形花柱性

heterotroph：有機（従属）栄養生物

heterozygosity：ヘテロ接合度、異種接合性、ヘテロ接合性

heterozygote genotype：ヘテロ接合体遺伝子型、異種接合体遺伝子型

heterozygous：異種接合体の、ヘテロ接合体の

heterozygous for the white allele：白色対立遺伝子のヘテロ接合

heterozygous individual：異種接合体の個体、対立遺伝子をヘテロで持つ個体、ヘテロ接合個体、ヘテロ接合体の個体

Hewitson's blue hairstreak butterfly：ウラミドリシジミ

hexagon：六角形

hexapoda：六脚亜門、六本脚

hh：ヘッジホッグ（hedgehog）

hibernaculum：（複 .-la）、冬眠場所、越冬生息場所〔ラテン語〕

hibernate：越冬する

hibernating larva：越冬幼虫

hibernation：冬眠、越冬

hibernation form：越冬形態、越冬型

hierarchical dominance：階層的優性

hierarchical F statistics：階層的 F 統計

hierarchical G test：階層的 G 検定、階層的対数尤度比検定

hierarchical likelihood ratio test：階層的尤度比検定

high dose-refuge strategy：高薬量／保護区戦略、高用量／保護区戦略

high elevation butterfly：高山蝶

high humidity：高湿度

high intensity：高強度

high level：高濃度

high magnification：高倍率、高拡大図

high performance liquid chromatography：高速液体クロマトグラフィー

high sound pressure：高音圧

high temperature：高温

high temperature sensitivity：高温感受性

high-molecular weight：高分子量

higher classification：高次分類

higher frequency：高周波数

higher-taxon richness：高次分類数

highland species：高地性種

highly polyploid branched nucleus：多倍数体分岐細胞核

highly unlikely generally：一般的にはあまりない、一般的にはあり得ない

highly-modified habitat：高度に改変された生息地
hill topping：ヒルトッピング、山頂占有性
hill-topping behavior：山頂占有行動
hillside：丘陵地帯
Himalayan type：ヒマラヤ型
hind leg：後脚(こうきゃく)、後肢(こうし)
hind thorax：後胸(こうきょう)
hindgut：後腸
hindwing (HW)：後翅(こうし)
hindwing coupling：後翅の結合部
hindwing margin：後翅縁
hindwing tail：尾状突起
HIPV：植食者が誘導する植物の揮発成分、植食者誘導性植物揮発性物質（Herbivore-Induced Plant Volatile）
histological cross section：組織学的断面
histology：組織学
histolysis：解離
historical biogeography：歴史(的)生物地理学
history：経緯
hitchhiker species：付着した外来種
hol-：完全-、全-〔ギリシャ語〕
Holarctic region：全北区(ぜんほっく)
holistic conceptual model：全体論的概念モデル
holo-：完全-、全-〔ギリシャ語〕
Holometabola：完全変態する昆虫類
holometabolous：完全変態の
holometaboly：完全変態
holotype：ホロタイプ、完模式標本、正基準標本
home range：行動圏、ホームレンジ
homeosis：ホメオシス、相同異質形成
homeothermic：定温の

homeotic gene：ホメオティック遺伝子、相同異質形成遺伝子
homeotic mutation：ホメオティック（突然）変異
homing habit：回帰性
homo-：同-〔ギリシャ語〕
homodynamic：同動態的、連続性繁殖動態的、ホモダイナミック、［注］休眠なしの生活環）
homoeisis：ホモエイシス
homogametic sex：同型性、同型配偶子をもつ性
homogenate：ホモジネート、組織粉砕懸濁液
homogeneity：等質性
homogenous subset of sample：サンプルの相同サブセット
homogeny：相同性
homolog of doublesex：両性遺伝子の相同体、両性の相同遺伝子、両性のホモログ遺伝子
homologous chromosome：相同染色体
homologous genetic pathway：相同遺伝経路
homologous linkage group：相同連鎖群
homologous marker：相同マーカー
homologous multichromosomal mimicry architecture：多相同染色体の擬態構成
homologous nerve branch：相同(的)神経枝
homologous structure：相同(的)構造、相同構造
homology：ホモロジー、相同
homology model：相同(性)モデル
homology modeling：ホモロジーモデリング、相同体モデル化

homology search：ホモロジー検索、相同性検索
homonym：同名異物、ホモニム、同名
homonymy：同名関係、同名状態
homozygote：ホモ接合体、同型接合体
homozygous for the yellow allele：黄色対立遺伝子のホモ接合
honey gland：蜜腺
honeybee：ミツバチ
honeydew：蜜、糖液、甘露
honorary authorship：名誉のオーサーシップ、名誉著者
Honshu population：本州個体群
hook：鉤爪（かぎづめ）
hooked：鉤爪状の
hooked hindwing tail：鉤爪状の尾状突起
Hopkins host selection principle：ホプキンスの寄主選択則
hopperburn：坪枯れ
horizontal distribution：水平分布
horizontal infection：水平感染
hormonal control：ホルモン制御、ホルモン調節、ホルモン支配
hormonal mechanism：ホルモン機構
hormone：ホルモン
hormone dynamics：ホルモン動態
hormone titer：ホルモン力価
horn：角状突起
horn length：角状突起長
horn type：角状突起の形態
horned head：つのがある頭部
hornet：スズメバチ
horticulturalist：園芸家
host：寄主、宿主（しゅくしゅ）
host affiliation：寄主の協力、寄主起源の認定

host arthropod：宿主節足動物
host cyanogenesis：寄主によるシアン発生、寄主によるシアン形成
host elongation factor：宿主由来の伸長因子、宿主伸長因子
host fitness：宿主の適応度
host generation：宿主世代
host plant：食樹、食草、寄主植物
host plant conspicuousness：寄主植物の目立ちやすさ、寄主植物の被視認性
host plant recognition：食草認識
host plant selection mechanism：食草選択機構
host plant species：寄主植物種
host population：宿主個体群、寄主個体群
host preference：寄主選好性
host preference shift：寄主選好性転換
host race：ホストレース、寄主品種、寄主系統
host range：寄主範囲
host shift：寄主転換
host specialization：食草特化適応
host specificity：宿主特異性
host tissue：宿主組織
host-marking pheromone：寄主マーキングフェロモン
host-plant specialization：寄主植物特異性
host-plant use：寄主植物利用
hostile：外敵に満ちた、相反する、敵対する
hostmarker：寄主マーカー、寄主標識
hot and humid forest：高温・湿潤な森林、高温多湿な森林
hot and humid period：高温湿潤期
house built on the sand：砂上の楼閣

house of card：砂上の楼閣
house spider：タナグモ
hovering briefly：一時的にホバリングして
Hox：ホメオティック遺伝子、相同異質形成遺伝子（Homeotic gene あるいは Homeobox gene）
Hox gene：ホメオティック遺伝子、相同異質形成遺伝子
HP：ヘアペンシル、毛束（HairPencil）
HPLC：高速液体クロマトグラフィー（High Performance Liquid Chromatography）
huge swathe：広大な帯状土
human prosperity：人間の繁栄
human welfare：人間の福祉
human-induced disturbance：人為的攪乱
humeral crossvein：肩横脈、h 脈
humeral lobe：肩葉
humeral vein：肩脈
humid condition：湿度条件
humid tropics：湿潤熱帯
humid zone：湿潤地帯
humidity：湿度
hummingbird hawk moth：ホウジャク、ホシホウジャク
hump-shaped diversity curve：こぶ状の多様性曲線
humus：腐葉土
hunter：採集者
HvirCR4：オオタバコガの仲間（*Helicoverpa virescens*）の化学受容体 4
hyaline spot：透明な斑点
hybrid：雑種、交雑
hybrid breakdown：雑種崩壊
hybrid dysfunction：雑種の機能不全
hybrid egg hatch：雑種卵の孵化
hybrid exchange of gene：交雑による遺伝子交換
hybrid genome：雑種のゲノム
hybrid inviability：雑種の生存不能、雑種致死、雑種死滅、雑種の生存力低下
hybrid male：雑種の雄
hybrid mating：雑種間の交配
hybrid rice：ハイブリッド米
hybrid speciation event：交雑による種分化イベント、交雑による種形成イベント、雑種種分化イベント
hybrid sterility：雑種不稔、雑種不妊性
hybrid viability experiment：雑種の生存力実験
hybrid wing：雑種の翅
hybrid zone：交雑帯
hybridization：交雑、種間交雑
hybridize：種間交雑
hybridizing species：交雑種
hydration：水和
hydrocarbon：炭化水素
hydrophilic protein：親水性タンパク質
Hymenoptera：ハチ目、膜翅目
hymenopteran insect：膜翅目の昆虫、ハチ目の昆虫
hypandrium：生殖下板
hyper-：超 -、過度の -〔ギリシャ語〕
hyperdiverse taxa：超多様な分類、極めて多様な分類
hyperfine structure：超微細構造
hypergeometric rarefaction curve：超幾何学的希薄化曲線
hypergeometric sampling distribution：超幾何学的サンプリング分布
hyperparasitism：過剰寄生
hyphen：ハイフン

hypothesis of age and area：時間と広がり説、年代領域説
hypothesis of neutral evolution：中立進化説
hypothesis of strong developmental constraint：強い発生的制約説
hypothesize：仮定する
hypothetical concept：仮説的概念、仮説上の概念
hypothetical example：仮想例、仮想事例
hypothetical, sample-based rarefaction curve：仮想的サンプル数に基づく希薄化曲線
HZG5：オオタバコガ（旧学名：*Heliothis zea*）の卵巣由来の樹立培養細胞株

i

IAs：侵略的外来種、特定外来生物（Invasive Alien species）
IBM：総合的生物多様性管理（Integrated Biodiversity Management）
IBR：昆虫行動制御剤、昆虫行動制御物質（Insect Behavior Regulator）
ice seeding：植氷
ice-inoculation avoidance：植氷凍結回避
ice-nucleating agent：氷晶核、氷核形成
iceplant：オオベンケイソウ
ichnotaxon：生痕化石タクソン
ICIPE：国際昆虫生理生態学センター（International Centre of Insect Physiology and Ecology）
iconographia insectorium Japonicorum colore naturali edita：原色日本昆虫図鑑〔ラテン語〕
ideal system：理想的なシステム、理想的体系
idealized spectrum：理想化されたスペクトル、理想的なスペクトル
identical cumulative number of species：同一の累積種数
identification：同定
identity：同定、識別
idiosyncratic：特異的な
if any：もしあれば
igneous rock：火成岩
IGR：昆虫成長制御物質、昆虫成長制御剤（Insect Growth Regulator）
illegal introduction：違法導入
Illumina technology：イルミナ技術
illuminant difference：照度差
illumination：照度
image-resolving eye：解像能力を有する眼
imaginal disc：成虫芽、成虫原基、成虫盤
imaginal myrmecophily：成虫の好蟻性
imago：成虫、イマーゴ〔ラテン語〕
imago form：成虫形態、成虫型
immature stage：幼虫期、未成熟期
immediate consequence：即座に現れる影響
immigrant：移入種
immigrant species：移入種
immigration：移入、移住して来ること
immobilization：不動化
immunity-related gene family：免疫系関連遺伝子ファミリー
immunohistochemical localization：免疫組織化学的局在
impair：減じる、弱める
imprinting：刷り込み、インプリンティング
in press：印刷中

in silico：インシリコ、シリコン内で、コンピュータを用いて〔ラテン語〕

in situ hybridization：インサイチュー・ハイブリダイゼーション法、「その細胞が由来する生物個体内の本来あるべき場所」での交雑実験、原位置標識法〔ラテン語〕

in the sense of：という意味での

in vitro：インビトロ、試験管内で、生体外で〔ラテン語〕

in vivo：インビボ、生体内で〔ラテン語〕

in-：内 -、中 -、反 -〔ラテン語〕

inability of eyespot：眼状紋の発育不全

inability of medial band：内側の縞状バンドの発育不全、中央部の縞状バンドの発育不全

inadvertent error：不慮の過誤

inappropriate name：不適切名

inbred：近親交配の、同系交配の

inbreeding：近親交配、インブリーディング、自殖、同系交配、近交

inbreeding depression：近交弱勢、近親交配弱勢

inbreeding species：近親交配種

incertae sedis：所属不明〔ラテン語〕

incidence light：入射光

incidence of diapause：休眠率、休眠の発生

incidence of pupal diapause：蛹休眠率

incipient species：発端種、初期種

incipient stage of speciation：種分化の初期段階

inclivous：内斜

inclusion-body disease：封入体症、封入体病

inclusive approach：包括的アプローチ

inclusive fitness：包括適応、包括適応度

income breeder：インカムブリーダー（産卵時の餌に依存する）

incoming female：飛来雌、侵入した雌

incoming virgin female：飛来してくる無交尾の雌

incommensurate area：不釣合いな面積

incompatibility：不和合性、不一致

incompatibility-inducing microbe：不和合性誘発微生物

incompatible cross：不和合性交配

incomplete metamorphosis：不完全変態

incorrect original spelling：不正な原綴(つづ)り

incorrect subsequent spelling：不正な後綴(つづ)り

incorrectly assess：誤認する、不正評価する

increasing stimulus intensity：増大する刺激強度

increasing titer：増加するタイター、増加する力価

incubator：恒温器、定温器、孵化器

incur：負う

indel：インデル、挿入欠失

indented：くぼんだ

independent locus：独立した遺伝子座

independent mating：自主的な交尾

independent observer：独立の観測機器

independent species：独立種

index：指標、索引

index fossil：標準（示準）化石

Indian leaf butterfly：コノハチョウ

indication：指示

indicator：指標

indigenous species：土着種（どちゃくしゅ）、

在来生物種
indirect defense：間接防衛
indirect effect：間接効果
indirect evidence：間接的証拠
indirect interaction web：間接相互作用網
indirect selection：間接的選択
indispensable：不可欠な
indistinct brown outer ring：不明瞭な褐色の外側の環
individual：個体
individual density：個体密度
individual difference：個体差
individual organism：生物個体
individual stem：単木
individual variation：個体変異、個体変動、個体多様性
individual-based accumulation curve：個体数に基づく累積曲線
individual-based curve：個体数に基づく曲線
individual-based dataset：個体数に基づくデータセット
individual-based protocol：個体数に基づくプロトコール
individual-based rarefaction：個体数に基づく希薄化
individual-based rarefaction formula：個体数に基づく希薄化の計算式
individual-based taxon-sampling curve：個体数に基づく分類サンプリングの曲線
individuals：個体数
Indo-Australian region：インド・オーストラリア区
indolequinone compound：インドールキノン化合物

indolic melanin：インドールメラニン
induced defense：誘導防衛
induction condition of autumn morph：秋型誘導条件
induction factor of summer form：夏型誘導因子
industrial melanism：工業黒化型、工業暗化
inevitable：避けられない
inevitably：必ず
infect：感染する
infected male：感染雄
infection density：感染密度
infection frequency：感染頻度
infectious disease：感染病
infectious germ：伝染病菌、感染性細菌
infectious parthenogenesis：感染性単為生殖
infectivity：感染性
inferior：下部の、下（方）の、下（か、した）〔ラテン語〕
inferiority：劣勢、劣等、不利
infiltrate：浸透する
inflorescence：花序
influence of microclimate：微気象の影響
influx species：流入した種
infra-：下 -〔ラテン語〕
infraorder：下目
infrared region：赤外域
infrared spectroscopy：赤外分光
infraspecific name：種よりも低位の学名
infrasubspecific：亜種よりも低位の
infrasubspecific name：亜種よりも低位の学名
infrasubspecific taxon：亜種よりも低位のタクソン

infrequent：めったに起こらない、まれな
infuscated：すす色の、黒ずんだ
infusion：注入
ingest：摂取する
ingredient：原材料
ingroup：イングループ、内集団
inhabit：生息する
inhabitant：生息生物
inheritable variation：遺伝的変異
inheritance：遺伝、継承
inhibiting factor：抑制要因、阻害要因
inhibition of summer-form induction：夏型の誘導阻害
inhibitory condition：抑制条件、阻害条件
inhospitable climate：住みにくかった気候
initial experiment：初回実験、初期実験
initiator of speciation：種分化の開始者
injection：注射、注入
Inka：インカ
innate character：本質的な性格
innate color preference：生得的な色選好性
inner edge：内縁
inner margin：内縁、後縁
inner membrane：内膜
inner membrane surface：内膜表面
inner surface：内面
innervate：刺激する、器官を刺激する、神経を刺激する
innervation：神経刺激伝達、神経支配
innocuous：無害の
innovative change：革新的な変化
inoculated SCP：植氷過冷却点
inoculation：植氷、接種
inoculative release：接種的放飼法
inorganic nitrogenous ion：無機態窒素イオン
insect：昆虫、インセクト
insect behavior：昆虫（の）行動
insect behavior regulator：昆虫行動制御剤、昆虫行動制御物質
insect carrier：保菌昆虫、媒介昆虫
insect chemoreception：昆虫化学受容
insect ecology：昆虫生態学
insect growth regulator：昆虫成長制御剤、昆虫成長制御物質
insect material：供試虫、供試昆虫
insect pest：害虫
insect pest management：害虫管理
insect physiology：昆虫生理学
insect saline：昆虫食塩水
insect tissue：昆虫組織
insect-umbellifer association：昆虫-セリ科の関係
Insecta：昆虫綱、昆虫類
insectarium：昆虫館
insectary：昆虫飼育場
insecticidal crystal protein：殺虫性結晶タンパク質（ICP）
insecticide：殺虫剤
insecticide resistance：殺虫剤抵抗性
insecticide resistance management：殺虫剤抵抗性管理
insectivore：食虫（性）動物
insectivorous bird：食虫性の鳥
insectivorous bird species：食虫性の鳥種
inseminated female：受精した雌、受精雌
inseparable：不可分
inset：差込み図、挿入図、インセット
insist：主張する
insolation hour：日照時間
insolubility：難溶性

instar：齢(れい)、○齢幼虫、令(れい)〔ラテン語〕
instigate：扇動する
intact：損なわれていないで、そのままの
intact forest：手つかずの林
integrated approach：統合的アプローチ
integrated biodiversity management：総合的生物多様性管理
integrated control：総合防除
integrated pest management：総合的害虫管理、総合防除
integument：外皮、皮膚(ひふ)、殻
intensification：強化
intensify：強める、増感する
intensity of diapause：休眠深度
intensity of irradiation：照射強度
intensity of light exposure：照射される光の強度
intensity of sound stimulus：音刺激の強度、音の刺激強度
intensity-response relationship：強度 - 応答関係
intensive agriculture：集約農業
intensive chasing：しつこい追跡飛翔、積極的な追飛
intensive sampling：集中(的)サンプリング、集約サンプリング、集中的抽出法
intentional introduction：意図的導入
inter-：間 -、相互に -〔ラテン語〕
inter-racial：品種間の、亜種間の
inter-specific territory：種間ナワバリ
interaction：交互作用
interaction of environmental factor：環境要因の相互作用
interaction strength：交互作用の強さ
interbreed：交配

intercellular distance：細胞間の間隔
intercross：相互交配、兄妹交配
interested amateur：愛好家
interfamily：科間
interfere：干渉する
interference competition：干渉型競争
interference RNA：RNA 干渉法
interfertile：異種交配できる
intergeneric cross：属間交雑
intergeneric hybrid：属間雑種
interglacial period：間氷期
Intergovernmental Panel on Climate Change：気候変動に関する政府間パネル
intergrade：中間的段階
intergrade population：中間型個体群
interindividual variation：個体間変異
interior region：内陸地方
intermediate：中間(ちゅうかん)
intermediate habitat：中間的な好適生息地
intermediate measure：中間的措置
intermediate morph：中間型
intermediate phenotype：中間表現型
intermediate photoperiod：中間的な日長
intermediate rate：中間的な率
intermediate temperature：中間的な温度、中間温度
intermediate type：中間型
intermediate-temperate：中間温帯(植生帯)
intermediates：中間型種
internal：内(ない)、内部の
internal body feature：内部構造
internal coincidence model：内的符合モデル
internal epidermal pouch：内部表皮の袋
internal lamella：内板

internal morphology：内部形態
internal reproductive organ：内部生殖器官
International Plant Protection Convention：国際植物防疫条約
International Rice Research Institute：国際イネ研究所、国際稲研究所
interpatch movement：パッチ間移動
interphase nucleus：間期細胞核
interpolate：内挿する
interpolated name：挿入名
interpolation：内挿、補間
interpopulational variation：個体群間変異
interracial contact zone：亜種間の接触帯
interracial difference：亜種間差、亜種間の差異
interracial hybrid zone：亜種間の交雑帯
intersexual defect：間性欠陥、間性障害
intersexual defect hypothesis：間性欠陥説、間性障害説
intersexual host trait：宿主の間性形質
intersexual phenotype：間性表現型
intersexual selection：異性間選択、雌雄選択
intersexuality：間性現象、間性
interspace：間腔、翅脈間隙
interspecies：種間
interspecific competition：種間競争
interspecific gene flow：種間の遺伝子流動
interspecific horizontal transfer：異種間の水平移動、異種間の水平伝播
interspecific hybridization：種間交雑
interspecific mating：種間の交配
interspecific variation：種間変異
intersperse：まき散らす
intertropical convergence zone：熱帯の赤道収斂域、熱帯収束帯

interval mapping：区間マッピング法
intervening larval diapause：休眠を回避した幼虫
intra-：内-、中-〔ラテン語〕
intrageneric competition：属内競争
intrageneric speciation of subgenus：属内の亜属分化
intragenomic conflict：ゲノム内闘争
intraguild predation：ギルド内捕食
intrasexual selection：同性内選択、性淘汰、同性内性選択
intraspecific competition：種内競争
intraspecific horizontal transfer：種内間の水平移動、種内間の水平伝播
intraspecific mimicry：種内擬態
intraspecific territory：種内ナワバリ
intraspecific variation：種内変異
intriguingly：興味をひくように
intrinsic：内因的、内在的
intrinsic barrier：固有障壁、内的障壁
intrinsic optimum temperature for development：内因的な発育最適温度
intrinsic property：内在的性質
intrinsic rate of increase：内的増殖率
intrinsic rate of natural increase：内的自然増加率
intrinsic reproductive power：内的繁殖力
introduce：導入する
introduced colony：移入されたコロニー
introduced plant：移入植物
introduced species：導入種、移入種
introduction：導入、移入、はじめに、序、序論
introgression：遺伝子侵入、遺伝子移入、移入交雑、遺伝子浸透

intron：イントロン
intruder：侵入個体
inundative release：大量放飼法
invade：侵入する
invader：侵入生物、侵入者
invalid：無効な
invalid name：無効名
invalid nomenclatural act：無効な命令法的行為
invaluable perspective：価値のない視点
invariant site：不変部位
invasion：侵入、侵入種
invasion and establishment of unfamiliar land：未経験の土地への侵入・定着
invasive：侵略的
invasive alien species：侵略的外来種、特定外来生物
invasive garden plant：侵入種の園芸植物
invasive species：侵入種
invasive species of concern：問題の侵入種
invasive species release：侵入種の野外放出
invasive weed seed：侵入雑草の種子
invasiveness：侵略性
inverse ratio：逆比
inversion：逆位、反転、インバージョン
invertebrate：無脊椎動物
investigated season：調査季節
investigation material：調査材料
investment：投資量、外被、外殻
investment in reproduction：繁殖への投資量
ionic：イオンの
IPCC：気候変動に関する政府間パネル（Intergovernmental Panel on Climate Change）

IPM：総合的害虫管理、総合防除（Integrated Pest Management）
IPPC：国際植物防疫条約（International Plant Protection Convention）
IR：赤外分光（Infrared spectroscopy）
iridescent：光沢の、虹色の、玉虫色の
iridoid：イリドイド
iridoid glycoside：イリドイドグルコシド、イリドイド配糖体
IRM：殺虫剤抵抗性管理（Insecticide Resistance Management）
irradiation：照射
irregular diapause-related phenomenon：不規則な休眠に関連する現象
irregular projection：異常な突起部、不規則な突起部、不整形な突起部
IRRI：国際イネ研究所、国際稲研究所（International Rice Research Institute）
irrigated land：灌漑（かんがい）地
irrigated meadow：灌漑された草地、灌漑草地
irrigation：灌漑
irruption：大発生、大繁殖
irruptive：急増した、大発生した、侵入した
ISH：インサイチュー・ハイブリダイゼーション法（*In Situ* Hybridization）
island community：島の群集
island-hopping：アイランドホッピング、島巡り
islands：島嶼（とうしょ）
isofemale line：単雌系統
isofemale offspring：単雌の子孫
isolated chromophore：孤立発色団、分離発色団
isolated habitat：隔離的な生息地

isolated population：孤立化した個体群
isolated species：孤立種
isolating mechanism：隔離機構
isolation：隔離、分離
isolation of distribution：分布の断絶
isoleucine-to-methionine substitution：イソロイシンからメチオニンへの置換
Isoptera：シロアリ目、等翅目
isotope：アイソトープ、同位体
isotropic light field：等方性光条件
isoxanthopterin：イソキサントプテリン
ITCZ：熱帯の赤道収斂域、熱帯収束帯（InterTropical Convergence Zone）
itero-：繰り返す-〔ラテン語〕
iteroparous：多数回繁殖
iteroparous colonizer：多回繁殖性の移住種
ithomiine butterfly：トンボマダラ科の蝶
ithomiine community：トンボマダラ科の蝶の群集
IUCN：国際自然保護連合（自然及び天然資源の保全に関する国際同盟）（International Union for Conservation of Nature and Natural Resources）

j

jacamar：キリハシ科の鳥
jack-of-all-trades：なんでも屋、よろず屋
jagged curve：ぎざぎざの曲線
jagged shape：のこぎり歯のような形
Janus Green B：ヤヌスグリーンB（ミトコンドリアの染色に用いられる色素）
Japan mainland：日本本土
Japanese：日本（産）の
Japanese butterfly：日本産蝶
Japanese name：和名
Japanese Society of Applied Entomology and Zoology：日本応用動物昆虫学会（JSAEZ）
Japanese type：日本型
jaw：大腮（たいさい、おおあご）、大顎（だいがく、おおあご）、口部
jerky flight：ぎくしゃくと飛ぶ
jet-black：漆黒色、まっ黒の
JH：幼若ホルモン（ようじゃくほるもん）、幼虫ホルモン（Juvenile Hormone）
Johnston's organ：ジョンストン器官
joining behavior：結合行動
JPP-NET：植物防疫情報総合ネットワーク（Japan Plant Protection general information NETwork system）
JSAEZ：日本応用動物昆虫学会（Japanese Society of Applied Entomology and Zoology）
judge：査読者
jugal lobe：翅垂
jugum：翅垂
julia butterfly：チャイロドクチョウ
jumping bean：セバスチアナの種（メキシコ産トウダイグサ科植物の種子内にいるガの幼虫の動きに伴って種子が踊り動く）
junction：接合部、結合部
junior homonym：新参同名
junior synonym：下位同物異名、新参異名
justified emendation：正当な修正名
juvenile hormone：幼若ホルモン（ようじゃくほるもん）、幼虫ホルモン
juvenoid：ジュベノイド
juxta (jx)：ユクスタ〔ラテン語〕、挿入器腹板

k

K locus：K 遺伝子座
K-pest：K 害虫（K 戦略をとる害虫）
K-selection：K- 淘汰、K 選択
K-species：K 種（K 戦略をとる種）
Ka/Ks：非同義置換と同義置換の比（Non-synonymous (Ka) and synonymous (Ks) nucleotide substitution ratio）
kairomone：カイロモン
Karner blue butterfly：メリッサミヤマシジミ
karyotype：核型
katydid：キリギリス
kb：キロ塩基、キロベース（塩基数の単位）
key：検索表
key factor：重要要因、変動主要因
key factor analysis：変動主要因分析
key pest：対象とする害虫、キーペスト、重要害虫
key species：鍵種
key stage：キーステージ
key stimulus：鍵刺激
key stone species：キーストーン種、中枢種
key word：キーワード
Khw：白色抑制遺伝子、キヌレニンヒドロキシラーゼ - ホワイト（Kynurenine hydroxylase-white）
kilobase：キロ塩基、キロベース（塩基数の単位）
kin：血縁
kin group：血縁群
kin recognition：血縁認識
kin selection：血縁選択
king of summer forest：夏の雑木林の帝王
kingdom：界（分類階級の「かい」）
kinship：血縁関係、血縁、近親関係
kynurenine：キヌレニン

l

L：明期（Light）
L opsin gene sequence：L（長波長型）オプシン遺伝子配列（Long-wavelength）
L photopigment：L 視物質、長波長感受性視物質
L photopigment lineage：L 視物質系統
L-sensitive photopigment：長波長感受性視物質
labial palp：下唇鬚（かしんしゅ）〔英語〕
labial palpus：下唇鬚（かしんしゅ）〔ラテン語〕
labial segment：下唇節
labile defensive lipid：不安定な防御脂質
labium：下唇（かしん）、下唇（かしん）〔ラテン語〕
laboratory condition：実験室条件
laboratory hybridization experiment：実験室での交雑実験
laboratory stock：研究室の貯蔵品、研究室ストック
laboratory study：基礎研究、実験室研究
labral：上唇の
labrum：（複．-ra）、上唇（じょうしん）〔ラテン語〕
lacewing：クサカゲロウ
lactic acetic orcein：乳酸酢酸オルセイン
lacuna：ラクナ、（骨や組織中の）小腔
lacy：レース状、レース編みの
lady beetle：テントウムシ
ladybug：テントウムシ

lamella：（複.-lae）、層板、薄片〔ラテン語〕
lamella antevaginalis：前膣ラメラ、前膣片
lamella postvaginalis：後膣ラメラ、後膣片
lamina：視葉板、板部、薄膜、葉片〔ラテン語〕
land management：土地管理
land-use pattern：土地利用パターン
landing：着地、着陸
landmark：陸標、目じるし
landscape：景観、地形
landscape disturbance：景観の攪乱
landscape fragmentation：景観の分断化、景観断片化
landscape scale：景観規模
landscape structure：景観構造
landscape-scale conservation：景観規模の保全
lapsus calami：書きまちがい〔ラテン語〕
large：大型
large copper butterfly：オオベニシジミ
large distance：長距離
large green-banded blue butterfly：タスキシジミ
large posterior eyespot：後方の大きな眼状紋
large skipper butterfly：コキマダラセセリ
large white butterfly：オオモンシロチョウ
large wood nymph butterfly：オオモンヒカゲ
large-scale rearrangement：大規模な再編成
larva：（複.-ae）、幼虫（ようちゅう）、若虫（わかむし）〔ラテン語〕
larval aggregation：幼虫の集合性
larval body：幼虫体
larval carnivore：幼虫の肉食性

larval development：幼虫発達、幼虫生育
larval diapause：幼虫休眠
larval diapause induction：幼虫の休眠誘起
larval food plant：幼虫の食草、幼虫の食餌植物
larval foodplant：幼虫の食草
larval imaginal disc：幼虫期の成虫原基
larval imaginal wing disc：幼虫期の翅成虫原基
larval infection：幼虫の感染
larval instar：幼虫の齢期
larval integument：幼虫皮膚
larval molt：幼虫脱皮、幼虫期の脱皮
larval mortality：幼虫死亡数、幼虫死亡率
larval myrmecoxeny：幼虫の客棲性
larval overwintering：幼虫越冬
larval performance：幼虫の発育、幼虫生存力、幼虫の成育
larval stage：幼虫期
larval survival：幼虫の生存率
larval transfer：移行齢期
larval web count：幼虫の巣単位での頭数
larval wing disc：幼虫期の翅原基
last author：最後の著者、統括著者
last instar larva：終齢幼虫
last-larval stadium：終齢幼虫期、幼虫の終齢期、終齢幼虫
late autumn：晩秋
late evening：深夜、遅い夜、夜分
late fifth instar：五齢後期
late male-killing：後期型雄殺し
late pupa：蛹後期
late season growing condition：シーズン後期の成育条件
latency：潜時、潜伏期間

latency of response : 反応潜時、応答遅延
later analysis : 事後分析、事後解析
later consequence : より後で現れる影響
lateral : 正中線により近い外側(がいそく)、側の、側方の
lateral inhibition : 隣接抑制、側抑制、側方抑制
lateral ocellus : 側単眼
lateral spine plate : 側域刺毛板
lateral-basking : 傾斜日光浴、閉翅日光浴、側面日光浴
latex flow : 乳液の流れ、ラテックスの流出
latex quality : 乳液の質、ラテックスの質
latex-bearing leaf : ラテックスを分泌する葉
Latin : ラテン語
Latin name : 学名
latinize : ラテン語化する
latitude : 緯度
latitudinal cline : 緯度クライン、緯度の連続変異
latitudinal gradient : 緯度勾配
lauraceous plant : クスノキ科植物
laurel-montane oak forest and relictual type : 照葉 - 山地カシ林遺存型
lay egg : 産卵する
LD : 明暗周期(Light-Dark)
lead compound : 鉛化合物
leading edge : 先端、先端部、前縁部
leaf : 葉
leaf eater : 葉食者
leaf mimic : 葉に擬態
leaf roller : 葉を巻いて、ハマキムシ
leaf shelter : 葉のシェルター
leaf tissue : 葉組織

leaf toughness : 葉の硬さ
leaf vein : 葉脈
leafwing : 葉翼、翼葉
Leafwings : フタオチョウ亜科
learning : 学習
least-squares fitting : 最小二乗法
lectotype : レクトタイプ、選定基準標本
leg : 脚(きゃく)、足(あし)、肢(し)
leg disc : 肢原基
leg segment : 脚節、脚部分
legally protected species : 法的保護種
legitimate interest : 合法的な関心
legitimate purpose : 正当な目的
lek : レック、集団求愛場、求愛集団
length of microtrich : 微毛長、微毛の長さ
lengthwise : 縦方向に
Lepidoptera : チョウ目、鱗翅目
lepidopteran insect : 鱗翅目の昆虫
lepidopterist : 蝶(蛾)の研究者
Lepidopterological Society of Japan : 日本鱗翅学会(LSJ)
lepidopterology : 鱗翅類学
lepidopteron : (複 . -ran)、鱗翅類〔ラテン語〕、チョウ・ガ類
lethal : 致死的、致死
lethal level : 致死(限界)レベル
lethal limit : 致死限界
leucine : ロイシン
leucopterin : ロイコプテリン
levana form : レヴァナ型(アカマダラ)
level of confidence, 0.05 : 信頼水準5%
level of enforcement : 施行水準
lie beneath : 真下に置かれる
life cycle : 生活環、ライフサイクル、生活史
life cycle characteristic : 生活環特性

life cycle pattern：生活環の型、生活史の型
life habit：生活様式
life history：生活史
life history strategy：生活史戦略
life pattern：生活型
life phenomenon：生活現象
life resource：生活資源
life span：生涯、寿命
life table：生命表
life type：生活型
life zone：生息帯、生活圏、生態区域、生息域
life-cycle polymorphism：生活環多型、生活史多型
life-history characteristic：生活史特性
life-history trait：生活史形質、生活史特性
ligand：リガンド
ligate：結紮(けっさつ)する、糸でくくる
ligation experiment：結紮(けっさつ)実験
light：明るさ
light condition：光条件
light environment：光環境、視環境
light intensity：照度
light interruption response：光中断反応
light irradiation：光照射
light micrograph：光学顕微鏡写真
light microscope：光学顕微鏡
light period：明期
light phase：明期、明相
light pulse：光パルス
light time：明期
light trap：ライトトラップ
light wavelength region：光の波長領域
light-dark cycle：明暗周期、明暗サイクル、LDサイクル
light-receptor substance：光受容体物質
light-sensitive chromophore：光感受性発色団
lightness constancy：明るさ恒常性
likelihood：尤度
likelihood approach：尤度アプローチ
likelihood ratio test：尤度比検定
limestone grassland：石灰岩の草地
limestone plateau：石灰岩台地
line of weakness：脆弱線
lineage：系列、系統、系譜
lineage ancestral：系統の祖先
lineage-specific expansion of chemosensory gene：化学感覚遺伝子の系統別の拡張
linear hierarchy：線形階層
linear ramp：リニアランプ、線形ランプ
linkage between cue and preference：刺激と選好の間の連鎖
linkage disequilibrium：連鎖不平衡
linkage group：連鎖群
linkage group assignation：連鎖群の割り当て
linkage group association：連鎖群関係
linkage mapping：連鎖地図、リンケージマッピング
linkage mapping study：リンケージマッピング研究、連鎖分析研究
linkage order：連鎖の順序
Linnaean tautonymy：リンネ式同語反復
Linnaeus：リンネウス(スウェーデンの自然科学者)、リンネ
Linnean system：リンネ式動植物分類(命名)法
linoleic acid：リノール酸
linolenic acid：リノレン酸
lipid：脂質、リピッド
lipid accumulation：脂質蓄積

lipocalin：リポカリン
lipophilic substance：親油性物質
lipophorin：リポフォリン
liquid：液体
liquid air：液体空気
liquid oxygen：液体酸素
List of Available Names in Zoology：動物学における適格名リスト
list of Japanese butterfly：日本産蝶類リスト
live specimen：生きた標本、生きている標本
livestock：家畜
livestock grazing：家畜の放牧
living ancestor：生きている祖先
living picked flower：生きたまま摘まれた花
lizard：トカゲ
LL：全照明、恒明（continuous Light あるいは Light-Light）
LMC：局所的配偶競争（Local Mate Competition）
lobe：でっぱり、短い突起、突起、葉片
lobed：葉状の
lobulus：（視）小葉
lobulus varginalis：腟小葉
local：局所的な
local adaptation：局所適応、局部順応
local butterfly lover：地方の蝶愛好家
local extinction：局所的な絶滅
local gene duplication：遺伝子の局所的な複製
local irradiation：局所照射
local irradiation experiment：局所照射実験
local irradiation of light：局所的光照射

local mate competition：局所的配偶競争
local mimicry polymorphism：局所擬態多型、地方擬態多型、地域擬態多型
local population：局所個体群、地域個体群、地域集団
local population dynamics：地域個体群動態
local population size：局所個体群サイズ
local population viability：局所個体群の生存力
local selection：局所選択、局部選択、その土地による選択
local specialist：局所スペシャリスト
local strain：地方系統、地域系統
local variation：地方変異、地域変動、地域多様性
local variety：地方種、在来種
locally adapted trait：局所適応した形質
located midway：中ほどにある
loci coding：遺伝子座コーディング
locomotor mimicry：動作擬態
locus：（複．-ci）、遺伝子座、座位、座、場所、位置〔ラテン語〕
locust：バッタ、イナゴ
LOD：対数オッズ、ロッド（Logarithm of odds）
log-odds ratio：対数オッズ比、LOR
logarithm of odds：対数オッズ
logged forest：伐採された森
logged habitat：伐採された生息地
logging：伐採
logistic regression model：ロジスティック回帰モデル
long axis：長軸
long day length：長い日長
long distance：長距離

long distance migration：長距離移動
long distance pheromone：長距離のフェロモン
long photoperiod：長日日長
long term：長期間、長期
long-day exposure：長日にさらすこと、長日処理
long-day photoperiod：長日日長
long-day plant：長日植物
long-day regimen：長日養生
long-day treatment：長日処理
long-day type：長日型
long-day type response：長日型反応
long-horn type：長角型
long-horned beetle：カミキリムシ
long-range signal：長距離シグナル
long-standing controversy：長年にわたる論争
long-standing question：長年の疑問
long-tailed blue butterfly：ウラナミシジミ
long-tailed skipper butterfly：アオネオナガセセリ
long-term control：長期間の防除
long-term diapause：長期休眠
long-term evolution：長期的進化、長期間の進化
long-term range expansion：長期間に渡る生息範囲の拡大
long-wavelength-sensitive cone：長波長感受性錐体
long-wavelength-sensitive photopigment：長波長感受性視物質
longevity：寿命、生涯、長生き、長命
longicorn beetle：カミキリムシ
longitudinal：縦（じゅう）
longitudinal section：縦断面

longitudinal vein：縦脈、縦走脈
Longwings：ドクチョウ亜科
loose linkage：疎性連鎖
loss of *Distal-less* expression：末端部のない遺伝子発現の消失
loss of eyespot：眼状紋の消失
loss of function：機能欠損、機能喪失、機能退化
loss of scale：鱗粉の脱落
loss of species diversity：種多様性の喪失
low altitude：低地
low temperature：低温
low temperature and short day：低温短日
low temperature phase：低温期
low temperature shock：低温ショック
low temperature treatment：低温処理
low temperature-sensitive stage：低温の感受期
low-frequency sound：低周波数音、低周波音
low-growing：背丈の低い
low-nitrogen diet：低窒素試料
low-temperature drying period：低温乾燥期
low-temperature period：低温期間
low-temperature stimulus：低温刺激
low-temperature tolerance：冷（低）温耐性
lower frequency sound：低周波数音
lower lamina：翅裏面
lower-bound estimate：下限推定、下限推定値、下限値の推定
lower-taxon--higher-taxon ratio：低次分類数 - 高次分類数の比
lowest intensity：最低強度
lowest temperature：最低気温
lowland：低地

lowland population：平地個体群
lowland species：平地性種
lowland tropical rainforest：低地熱帯雨林
LSJ：日本鱗翅学会（Lepidopterological Society of Japan）
Luehdorfia line：ルードルフィア線
luminosity：照度
luna moth：オオミズアオ
lunule：半月紋、三日月形
lusting of ecdysone：エクダイソン欠除
lutein：ルテイン
lutexin：ルテキシン
luxuriant green foliage：繁茂している緑色の葉
lycaenid larva：シジミチョウ科の幼虫
lycaenid-ant association：シジミチョウとアリとの関係

m

m-species-list curve：m種リスト曲線
m-species-list method：m種リスト法
MacClade：マッククレイド（最大節約法に基づいて系統推定を行うソフトウェアの名称）
macerate：液体に浸して柔らかくする
macro lens：接写レンズ、マクロレンズ
macroevolution：大進化
macroinvertebrate：大型無脊椎動物
macrolepidoptera：大型チョウ目、大型鱗翅類
macropter：長翅型
macropterism：長翅
macropterous form：長翅型
macroscopic phenotype：巨視的な表現型
macular：黄斑の
Madagascan sunset moth：ニシキオオツバメガ
magnetic bead：磁性ビース
magnetic compass：磁気コンパス
magnifying glass：虫眼鏡
magnitude of winter severeness：冬期の厳しさの度合い
mainland community：本土の群集
maintenance of diversity：多様性の維持
maintenance of race：種族維持、系統維持
major barrier：大きな障壁
major ecosystem：大生態系
major gene：主遺伝子、主働遺伝子
majority：大部分、大多数
majority-rule consensus tree：多数決原理総意樹、多数決合意樹、多数合意樹、多数派支配型コンセンサス樹
malachite butterfly：ミドリタテハ
maladaptation：不適応
Malayan type：マレー型
male：雄（おす）、♂、オス
male clasper：雄の把握器
male color preference：雄の色彩選好
male courtship preference：雄の求愛選好
male embryo：雄（の）胚
male genitalia：雄性生殖器
male genotype：雄の遺伝子型
male mate choice：雄の配偶者選択、雄の配偶者選好性
male mate-locating behavior：雄の雌探し行動
male mating behavior：雄の配偶行動
male model：雄モデル
male phenotype：雄の表現型
male wing odour：雄の翅から発生する匂い
male-biased：雄に偏った、雄偏重

male-heterogametic sex chromosome constitution：雄ヘテロ型性染色体構成

male-informative：雄 - 有用な情報付き

male-killing：雄殺し

male-like and female-like phenotype：雄と雌の特徴とを合わせもつ表現型

male-like genital trait：雄の特徴を持つ生殖器表現形質

male-like reproductive organ：雄の特徴を持つ生殖器官

male-like wing trait：雄の特徴を持つ翅表現形質

male-specific embryonic mortality：雄特有の胚死亡率

male-specific molecular mechanism：雄に特有の分子機構

male-specific organ：雄特有の器官

male-transferred anti-aphrodisiac：雄から引き渡された抗催淫物質

malfunction：機能不全

mallow：ゼニアオイ

Malpighian tube：マルピーギ管

Malpighian tubule：マルピーギ管

Malpighian tubule cell：マルピーギ管細胞

mammal：哺乳類

managed area：管理区域

managed relocation：人為的な生息地移動

mandatory change：強制変更

mandible：大腮（たいさい、おおあご）、大顎（だいがく、おおあご）

mangrove：マングローブ

manifested：明白な、現れる

Mann-Whitney U-test：マン - ホイットニーの U 検定

manoeuvrability：操作能力、操縦性

mantid：カマキリ

Mantodea：カマキリ目、蟷螂目

Mantophasmatodea：カカトアルキ目、踵行目

manufacture：生産する

manuscript：草稿

mapping：マッピング

marginal：外縁部、外縁部帯

marginal band：外縁の縞状バンド

marginal eyespot：外縁部の眼状紋

marginal facies：縁辺相

marginal habitat：分布辺縁生息地、周ределen生息地、辺縁の生息地

marjoram：ハナハッカ、マヨラナ（シソ科の低木）

mark and recapture statistics：標識再捕獲統計、標識再捕獲統計学

mark and recapture trial：標識再捕実験

mark and release research：リリースするマーキング調査

mark-recapture：標識再捕獲法、標識再捕法

mark-recapture method：標識再捕獲法、標識再捕法

mark-release-recapture study：標識 - 放出 - 再捕獲の研究

marked difference：著しい差、顕著な相違

markedly lower rate of survival：著しく低い生存率

marker：マーカー、標識物質

marker locus：マーカー遺伝子座

marking：斑紋、模様

marking pattern of wing：翅斑紋パターン

marking pre-pattern：斑紋プレパターン

marking variation：斑紋変異

marsh：沼地、湿地
marsh fritillary butterfly：チョウセンヒョウモンモドキ
masquerade：みせかけ、仮装、なりすまし
mass capture：大量捕獲物
mass extinction：大量絶滅
mass spectrograph：質量分析計
master gene：マスター遺伝子
mate assortatively：同類的に交配する
mate choice：配偶者選択
mate choice experiment：配偶者選択実験
mate competition：配偶競争、配偶者競争、配偶者争奪戦
mate preference cue：配偶者選好の刺激
mate randomly in nature：自然ではランダムに交尾する、自然無作為交配
mate recognition：配偶者認知、配偶者認識
mate recognition signal：配偶者認知シグナル
mate refusal behavior：交尾拒否行動
mate seeking territory：探雌のための縄ばり
mate-locating behavior：雌探し行動、配偶者位置探索行動
mate-locating strategy：配偶者位置探索戦略
mate-refusal posture：交尾拒否姿勢
mate-searching behavior：探雌のための行動
mate-seeking response：探雌反応
mate-selection：配偶者選択
material and method：材料と方法
maternal age：母年齢
maternal effect：親世代の影響、母性効果
maternal species：母親種
maternal-form effect：母体型の影響

mathematical expression：数学的表現
mating：交尾(こうび)、交配
mating asymmetry：交尾の非対称性
mating attractiveness：交尾誘引性
mating behavior：配偶行動、交尾行動
mating between the two species：種間交配（種）
mating discrimination：交尾識別
mating pair：配偶ペア、交尾対
mating plug：交尾栓
mating probability：交尾確率
mating propensity：交尾傾向
mating strategy：交尾戦略
mating success：交尾成功率
mating system：配偶システム、配偶様式
mating territory：交尾テリトリー、雌を見つけるためのナワバリ、交尾ナワバリ
mating tube：交尾管
mating within species：種内交配（種）
matrix habitat：マトリックス生息地
maturation：成熟
maturation period：成熟期
mature egg：成熟卵
mature grassland：十分に成長した草地
mature oocyte：成熟卵母細胞
mature ovary：成熟した卵巣
mature ovum：成熟卵
maxilla：（複．-ae)、小腮(しょうさい、こあご）、小顎(しょうがく、こあご）〔ラテン語〕
maxillary galea：小腮外葉
maxillary lobe：小腮粒状体
maxillary palp：小腮鬚(しょうさいしゅ)、小顎鬚(しょうがくしゅ、こあごひげ）〔英語〕
maxillary palpus：小腮鬚、小顎鬚〔ラテン語〕
maxillary segment：小腮節

maximally dominant：最大限に優勢な

maximum and minimum thermometer：最高最低温度計

maximum likelihood method：最尤法、ML法

maximum log-likelihood：最大対数尤度

maximum parsimony：最大節約法

maximum parsimony method：最節約法、最大節約法、MP法

maximum-likelihood ancestral state reconstruction：最尤法による祖先状態再構成

maximum-likelihood phylogeny：最尤推定系統樹

maximum-parsimony analysis：最節約解析

maximum-parsimony ancestral state reconstruction：最節約法による祖先状態再構成

McDonald-Kreitman test：マクドナルド-クレイトマンテスト

meadow：牧草地

meadow brown butterfly：マキバジャノメ

meal：食事

mean：平均

mean generation time：平均世代時間

mean minimum temperature：平均最低気温

mean number of accumulated individual：平均累積個体数

mean number of stem per quadrat：方形区当りの平均樹幹数

mean richness value：平均種数値

mean supercooling point：平均過冷却点

MEAs：多国間環境協定（Multilateral Environmental Agreements）

measure of diversity：多様性指数

measure of patchiness：パッチネス指数

measure of sampling intensity：サンプリング強度の指数

measuring amplifier：測定増幅器、計測用増幅器

mechanical control：機械的防除

mechanical isolation：機械的隔離

mechanical sense：機械(的)感覚

mechanism of emergence：出現機構

mechanism of flight orientation：飛行方向の機構

mechanism of voltinism change：化性変化の機構

mechanoreceptive property：機械感覚特性

mechanoreceptor：機械感覚器、機械受容器

mechanosensory neuron：機械刺激ニューロン、機械刺激神経細胞

meconium：蛹便

Mecoptera：シリアゲムシ目、長翅目

Medea factor：MEDEA因子（Maternal-effect dominant embryonic arrest）、母性効果優勢胚発育停止因子、メディア因子（利己的な遺伝因子の一種）

media：中脈、M脈

medial：内側(ないそく)、正中線により近い(相対位置)、中央帯、中央の

medial ablation：内側部位の切除

medial band：内側の縞状バンド

medial crossvein：中横脈、m脈

medial microcautery：内側部位の微細焼灼

medial-cubital crossvein：中肘横脈、m-cu脈

medial-lateral axis：内外軸、ML軸

median : 中央帯、中央の、正中(せいちゅう)、中央値、メディアン、メジアン(【統計学】)
median line : 正中線
median plane : 正中断面
median section : 正中断面
median surface : 正中断面
median threshold : 中央閾(いき)値、メディアン閾値
median value : メディアン値、中央値
median vein : 中脈(ちゅうみゃく)、M 脈
mediate : 媒介する、調節する、媒体となる
mediated contact : 介在する接触
medical freezer : 医療用冷凍庫
medio- : 中央 -、内側 - 〔ラテン語〕
medio-cubital crossvein : 中肘横脈、m-cu 脈
medio-lateral axis : 内外軸、ML 軸
mediobasal : 内側基部
mediodiscal : 内側中央帯
mediolateral axis : 内外軸、ML 軸
Mediterranean plant : 地中海地域の植物、地中海植物
medius : 中脈
medulla : 視髄、髄層
mega pixel : メガピクセル
megascopic : 肉眼
meiosis : 減数分裂
meiotic drive : マイオティックドライブ、減数分裂分離ひずみ
melanic : 黒化の
melanic form : 黒化型、黒色型、暗色型
melanic pigment : メラニン色素
melanic scale : 黒化鱗粉
melanin : メラニン
melanized warm-season phenotype : 暖候期の黒化表現型
membrane : 膜
membrane protein : 膜タンパク質
membranous cuticle : 膜性表皮、膜質表皮層
membranous structure : 膜状の構造、膜構造
Mendelian factor : メンデル因子
meral suture : 頭基節縫合線
mercury vapour lamp : 水銀灯
mes- : 中 - 〔ギリシャ語〕
mesh : 網
meso- : 中 - 〔ギリシャ語〕
mesonotum : 中胸背板
mesoscutellum : 中胸小楯板(ちゅうきょうしょうじゅんばん)
mesoscutum : 中胸楯板、中胸盾板
mesosoma : 中体節
mesothoracic wing base : 中胸翅基部
mesothorax : 中胸(ちゅうきょう)
met- : 後 - 〔ギリシャ語〕
meta- : 後 - 〔ギリシャ語〕
meta-species : メタ種
metabolic : 代謝 (上) の
metabolic activity : 代謝活動、代謝活性
metabolic detoxification : 代謝 (性) 解毒
metabolic rate : 代謝速度
metabolism : 代謝、物質代謝
metallic : 金属色の、金属光沢の
Metalmark : シジミタテハ科の蝶の総称
metamorphic rock : 変成岩
metamorphic stage : 変態期
metamorphose : 変態する
metamorphosis : 変態
metanotum : 後胸背板
metapopulation : メタ個体群

metapopulation biology：メタ個体群生態学

metapopulation dynamics：メタ個体群動態

metapopulation dynamics analysis：メタ個体群動態解析

metapopulation dynamics model：メタ個体群動態モデル

metapopulation effect：メタ個体群効果

metapopulation viability：メタ個体群の生存能力、メタ個体群生存率、メタ個体群の存続性

metasoma：後体節

metathorax：後胸(こうきょう)

Metazoa：後生動物

methanol-acetic acid：メタノール酢酸

methionine：メチオニン

methionine-rich storage protein：メチオニン型貯蔵タンパク質

method of comparative morphology：比較形態学的方法

methodological advance：方法論的進歩

methyl salicylate：サルチル酸メチル

methylalkylpyrazine：メチルアルキルピラジン

MFO：混合機能酸化酵素（Mixed Function Oxydase）

Miami blue butterfly：マイアミミドリシジミ

miconium：蛹便、ミコニウム

micro structure：微細構造

microbial assemblage：微生物集団、微生物群集

microbial community：微生物群集

microbial control：微生物的防除

microbial insecticide：微生物的殺虫剤

microcautery：微細焼灼

microcentrifuge tube：微量遠心チューブ、マイクロ遠心チューブ、微量遠心管

microclimate：微気象

microclimatic constrain：微気象制約条件

microcosm：ミクロコスモス、小宇宙、微小生態系

microevolution：小進化

microhabitat：微生息場所、マイクロハビタット

microinjection：微小注射、マイクロインジェクション

microlepidoptera：小型鱗翅目、小型鱗翅類、小蛾類

micronodule：微小結節

microorganism：微生物

microphone：マイクロホン

micropterous form：短翅型

micropylar：精孔の、卵門(らんもん)の

micropyle：精孔、卵門

microsatellite：マイクロサテライト

microsatellite marker：マイクロサテライトマーカー

microscope slide：顕微鏡用スライドグラス、顕微鏡用スライド

microscopic endosymbiont：微視的な細胞内共生生物

microscopic observation：顕微鏡観察

microscopic specimen：微視的な標本、顕微鏡標本

microsite：微高地、微環境、微地形、マイクロサイト

microsporidiosis：微胞子虫病

microsporidium：(複.-ia)、微胞子虫

microtrich：微毛

microvillus：(複.-li)、微絨毛

mid leg：中脚(ちゅうきゃく)、中肢(ちゅうし)

mid- : 中 - 〔英語〕
mid-autumn : 秋の半ば
mid-final instar larva : 中終齢幼虫
Mid-Tertiary : 第三紀中葉、第三紀中期
middle : 中央の
middle branch : 中枝
middle larval stage : 幼虫期の中頃
middle leg : 中脚(ちゅうきゃく)、中肢(ちゅうし)
middle-wavelength-sensitive cone pigment : 中波長感受性錐体視物質
middle-wavelength-sensitive photopigment : 中波長感受性視物質
middorsal : 背部中央
midgut : 中腸
midline : 中線、正中線
midline of symmetry system : 相称系の正中線
midnight : 真夜中
midrib : 葉脈
midsummer : 真夏
midventral : 腹中線、中腹側
migrant : 移動性種
migrant skipper butterfly : イチモンジセセリ
migrant species : 移動性種
migrate : 渡りをする
migration : 移動、渡り、回遊、移住、分散
migration factor : 移動指数
migration group : 移動群
migration rate : 移動率
migration syndrome : 移動形質群
migration-colonization syndrome : 移動 - 定着形質群
migratory : 移動性の

migratory butterfly : 移動性の蝶
migratory habit : 移動性、移動習性
migratory species : 移動性種
Milbert's tortoiseshell butterfly : ヤンキーコヒオドシ、アメリカコヒオドシ
milkweed : トウワタ
Milkweed Butterflies : マダラチョウ亜科、マダラチョウ科
milkweed butterfly : オオカバマダラ
million : 100万
mimeographing : 謄写版印刷
mimesis : 隠蔽擬態、隠蔽的擬態(いんぺいてきぎたい)、擬態、ミメシス
mimetic association : 擬態関係
mimetic convergence : 擬態の収斂
mimetic group : 擬態群、擬態グループ
mimetic insect : 擬態昆虫
mimetic lineage : 擬態系統
mimetic pattern : 擬態パターン
mimetic polymorphism : 擬態多型
mimetic pressure : 擬態圧
mimic : 擬態しているもの、擬態する、擬態種、ミミック
Mimic-Whites : トンボシロチョウ亜科
mimicry : 擬態、標識擬態、標識的擬態、ミミクリー
mimicry association : 擬態関係
mimicry evolution : 擬態進化
mimicry locus : 擬態遺伝子座
mimicry pattern : 擬態のパターン
mimicry polymorphism : 擬態多型
mimicry ring : 擬態リング
mimicry variation : 擬態変異
mineral : ミネラル、無機物
minimal interference : 最小限の干渉
minimum temperature : 最低気温

minimum threshold：最小閾(いき)値
Ministry of the Environment：環境省(日本)
Miocene series：中新統、中新世
miriamide：ミリアミド
mirrored tapetum：鏡型反射層板
misapply：誤適用する
misidentify：誤同定する
misleading result：誤った結果
missing：欠失、欠損、欠如、欠落
missing haplotype：欠損ハプロタイプ、欠失単模式種
Mission blue butterfly：ミッションミドリシジミ
mist：霧
misunderstanding：誤認
mite：ダニ
mitigation：影響緩和、環境緩和
mitigation measure：影響緩和措置
mitochondrial DNA：ミトコンドリアDNA
mitochondrial gene：ミトコンドリア遺伝子
mitochondrial gene encoding：ミトコンドリア遺伝子の符号化
mitochondrial gene sequence：ミトコンドリア遺伝子配列
mitotic frequency：細胞分裂頻度
mixed breed：雑種、混交一種
mixed function oxydase：混合機能酸化酵素
mixed strategy：混合戦略
mixture：混在
MK test：MKテスト、マクドナルド - クレイトマンテスト(McDonald-Kreitman test)
ML axis：内外軸、ML軸(Medial-Lateral axis)
mobile：移動性の

mobile species：移動性種
mobility：移動性
model：モデル
model of bovine rhodopsin：ウシロドプシンモデル
model-predicted average hostplant preference：モデルで予想した平均的な寄主植物選好性
modeling clay：模型用粘土
modern human：現代人、現生人
modern synthesis：現代の総合、現代的統合
modification-rescue model：修飾 - 救済モデル
modified sperm：修飾された精子、精子修飾
modifier：変更遺伝子、修飾因子、修飾遺伝子
modulating effect：調節効果
moist paper towel：湿ったペーパータオル
moisture：湿気
molecular component：分子成分
molecular evolution：分子(的)進化
molecular evolutionary analysis：分子進化(学)的解析
molecular evolutionary approach：分子進化(学)的研究法
molecular genetics：分子遺伝学
molecular marker：分子マーカー
molecular mechanism：分子機構
molecular phylogenetic analysis：分子系統学的解析
molecular phylogenetical analysis：分子系統学的解析
molecular phylogenetical point of view：分子系統学的観点

molecular phylogenetical study：分子系統学的研究
molecular phylogenetical tree：分子系統樹
molecular phylogenetics：分子系統(学)
molecular phylogeny：分子系統(学)
molecule：分子
mollusk：軟体動物
molt：脱皮(する)
molt instar stage：脱皮齢期
molting：眠、脱皮(だっぴ)
molting hormone：脱皮ホルモン
moltinism：眠性
monandrous female：単数回交尾の雌
monandry：単数回交尾
monarch butterfly：オオカバマダラ
monitoring：モニタリング
monitoring number and species：個体種数調査
monkey flower：ミゾホオズキ
mono-：1-、単-〔ギリシャ語〕
monobasic taxa：一つしか含まれない分類
monocarpic perennial plant：一回結実型多年生植物
monocarpy：一回繁殖性
monocotyledon：単子葉植物
monogamy：一夫一妻、一夫一婦、単婚
monogenesis：単為生殖(たんいせいしょく)
monogyny：単婚、単女王性
monolayer sheet：単分子膜細胞層
monomorphic：単一型、単型
monooxygenase：一原子酸素添加酵素
monophagous：単食性の、単食性、単食
monophagy：単食性
monophyletic group：単系統群
monophyletic sister group：単系統姉妹群
monophyletic species concept：単系統種概念
monophyly：単系統、単系統性
monostemma-dominated neuron：単一単眼優性ニューロン
monotrysia：単門亜目
monotypic genus：一属一種の属、単一種属
monotypy：単型
monsoon forest：雨緑林
montane：山地性の
montane oak forest：山地カシ林
montane population：山地(性)個体群
monthly average precipitation：月平均降水量
monthly total precipitation：月降水量
morph：モルフ、異形態
morpho butterfly：モルフォチョウ
morphogen：モルフォゲン(位置情報物質)
morphogen gradient：モルフォゲン勾配、形原勾配
morphogen model：形原モデル
morphogenesis：形態形成
morphogenetic：形態形成の、形態形成的、形態発生の
morphogenetic hormone：形態形成ホルモン
morphological adaptation：形態的適応
morphological analysis：形態学的解析
morphological and physiological characteristic：形態的・生理的特性、形態的・生理的特徴
morphological change：形態変化
morphological character：形態(的)形質、形態的特徴
morphological characteristic：形態(的)形質

morphological classification：形態(学)的分類
morphological datum：(複.-ta)、形態(学的)データ
morphological difference：形態的差異
morphological discontinuity：形態的不連続性
morphological distinction：形態的区別
morphological divergence：形態的分化
morphological diversification：形態的多様性
morphological diversity：形態的多様性
morphological evolution：形態的進化
morphological feature：形態(学)的特徴
morphological pattern：形態的パターン
morphological phylogenetics：形態系統学
morphological resemblance：形態(学)的な類似
morphological speciation：形態的分化
morphological species concept：形態学的種概念
morphological study：形態(学)的研究
morphological trait：形態(的)形質
morphological tree：形態系統樹
morphological variation：形態的変異
morphological, physiological and behavioral study：形態学的・生理学的・行動学的研究
morphologically abnormal adult insect：形態的に異常な成虫
morphology：形態学、形態
morphology measurement：形態測定
Morphos：モルフォチョウ亜科
morphospecies：形態(学的)種
mortality：死亡率
mosaic pattern：モザイク状

most order：最上位の遺伝子配列順序
most parsimonious model：最節約モデル、MPモデル
moth：蛾
moth-butterflies, the (superfamily Hedyloidea)：シャクガモドキ上科
moth-pollinated flower：蛾媒花
mother cell：母細胞
mother plant：母植物、親株、母株
motif：モチーフ
motile：自分で動くことができる、自動性の
motion vision：動態覚、動体視覚、動体視
motmot：ハチクイモドキ(科の鳥)
motor activity：運動活動、運動活性
moulting：脱皮〔英語〕
moultinism：眠性
mountain：山地
mountain ecotype：山地生態型
mountain population：山地性個体群
mountainous area：山岳地帯
mountainous species：山岳性種
mounted body：固定した胴体
mounting：展翅(てんし)
mounting board：展翅板
mounting fresh specimen：生展翅
mounting relaxed specimen：軟化展翅
mourning cloak butterfly：キベリタテハ
mouse：(複.mice)、ネズミ、マウス
mouth：口、口器
mouth parts：口器(こうき)
mouthparts：口器
movement：移動、運動
movement pattern：行動パターン

moving trace：行動トレース

MrBayes：ベイズ法に基づいて系統樹探索を行うソフトウェアの名称

Ms：質量分析計（Mass spectrograph）

mtDNA：ミトコンドリア DNA

mucilage sac：粘液のう

mud：泥

mud puddle：泥の水溜まり

mud-puddling behavior：泥水の吸水行動

muddy ground：ぬかるみ

Mullerian mimicry：ミュラー型擬態、ミュラー擬態

multi-：多-〔ラテン語〕

multi-brooded：多化性の

multidimensional scaling analysis：多次元尺度分析、多次元尺度解析

Multilateral Environmental Agreements：多国間環境協定

multilocus color pattern architecture：多遺伝子座で構成されるカラーパターン構成

multilocus genotypic difference：多遺伝子座で構成される遺伝子型の相違

multiple authorship：多数著者

multiple brood：多化

multiple invasion：多重侵入

multiple original spelling：複数原綴（つづ）り

multiple-site nester：複数箇所営巣性種

multiply：繁殖する

multispecies community：多種群集

multistemmata-dominated neuron：複数単眼優性ニューロン

multivoltine：多化性

multivoltine group：多化性集団

multivoltine population：多化性集団

murexide reaction：ムレキシド反応

muscle receptor：筋受容器

muscle-filled thorax：筋肉で満たされた胸部

musculature：筋肉組織、筋組織

museum specimen：博物館の標本

mustard oil glycoside：マスタード油グリコシド、マスタード油配糖体

mutagenize：突然変異を起こさせる

mutant：突然変異種、突然変異体

mutant eyespot phenotype：突然変異型眼状紋表現型

mutant phenotype：突然変異体表現型

mutation：突然変異、変異

mutation rate：変異速度、変異率、突然変異率

mute：音を出さない

mute butterfly：音を出さない蝶

mutual pursuit：相互の追跡

mutualism：相利共生

mutualist ant：相利共生者のアリ

mutualistic：相利共生的、相利型

mutualistic interaction：相利共生的相互作用

mya：100万年前（million years ago）

myrmecophilous insect：好蟻性昆虫

myrmecophilous organ：好蟻性器官

myrmecophily：好蟻性、アリとの共生

myrmecophyte：アリ植物（アリに食物や巣を提供している植物）

myrmecoxenous：客棲性（「蟻の客として迎え入れられて蟻と一緒に蟻の巣に棲む性質」のこと）〔ギリシャ語〕

myrmecoxenous relationship：客棲性的共生関係

myxomatosis：粘液腫病

n

N-containing compound：窒素含有物
N-β-alanyldopamine：N-β-アラニルドーパミン
nAChR：ニコチン性アセチルコリン受容体（nicotinic Acetylcholine Receptor）
NADH：ニコチンアミドアデニンジヌクレオチドの還元型（Nicotinamide Adenine Dinucleotide Hydrate）
NADH dehydrogenase subunit 5：NADHデヒドロゲナーゼサブユニット5
name：名称
name of a family：科名
name of a genus：属名
name of a species：種名
name of a subfamily：亜科名
name of a subgenus：亜属名
name of a subspecies：亜種名
name of a superfamily：上科名
name of a tribe：族名
name-bearing type：担名タイプ
named specimen：標本の命名
namely：すなわち、つまり
naming system：命名法
Nanda-Hamner protocol：ナンダ-ハムナープロトコール、ナンダ-ハムナー実験
narrow contact zone：狭接触帯、狭い接触域
narrow distribution：狭分布
narrow geographic area：地理的に狭い地域
narrow zone：狭い領域
narrow-distributed species：狭分布種
narrowly distributed species：狭分布種、狭域分布種
natality：出生率
national annual index of abundance：国レベルの豊富さの年次指数
national butterfly：国蝶
national butterfly of Japan：日本の国蝶
national park：国立公園
native element：自生的な要素
native garden plant：在来の園芸植物
native habitat：本来の生息地
native plant：土着植物
native species：在来種、土着種、在来種、固有種、自生種
native wildlife：在来野生生物
native woodland：自然のままの森林地帯
natural：自然の
natural barrier：自然の障壁
natural character：本来的性格
natural community：自然群集
natural competitor：自然の競争者
natural condition：自然条件
natural daylength：自然日長
natural disaster：自然災害
natural distribution of species：種の自然分布
natural enemy：天敵
natural environment：自然環境
natural fragmentation：自然な断片化
natural granitic outcrop：自然な花崗岩の露出部
natural group：自然群
natural habitat：自然（の）生息地
natural history：自然史
natural hybrid zone：自然交雑帯
natural incidence：自然発生、自然発生率
natural item：自然のもの、自然の代物

natural light：自然光
natural monument：天然記念物
natural period：固有周期
natural predator：自然の捕食者
natural range：自然分布域
natural resting position：自然な形の静止姿勢
natural seasonal variation：自然界の季節変異
natural selection：自然選択、自然淘汰
natural selection theory：自然選択説
natural vegetation：自然植生
natural vegetation fragment：自然植生の分断化
naturalization：帰化
naturalized：帰化した
naturalized organism：帰化生物
naturalized species：帰化種
naturally polymorphic race：生来の多型系統
naturally-occurring population：自然発生個体群
nature：自然（界）
nature conservation：自然保護、自然環境保全
nature of predation, the：捕食の本性
nature reserve：自然保護区
ND5：NADH デヒドロゲナーゼサブユニット5（NADH Dehydrogenase subunit 5）
Neanderthal：ネアンデルタール人
near-identical color-pattern mosaic：ほぼ同一なカラーパターンモザイク
near-isogenic line：近似同質遺伝子系統
near-perfect local mimetic convergence：ほぼ完全な擬態の局所的な収束

nearby natural habitat：近くの自然生息地
Nearctic region：新北区（しんほっく）
nearest relative：最寄りの近縁
nearly alike：大同小異
neck：頸部、首
nectar：蜜、花蜜〔ラテン語〕
nectar corridor：花蜜回廊
nectar guide：蜜標
nectar plant：吸蜜植物
nectar resource plant：吸蜜（源）植物
nectar source：吸蜜源
nectar-feeding insect：蜜を摂取する昆虫
nectaring：吸蜜
nectaring source：吸蜜源
nectarivore：蜜食動物
negative binomial：負の二項（分布）
negative feedback：負のフィードバック
Nei's D：ネイの遺伝（学）的距離
neighbor-joining：近隣結合法、NJ 法
neighbor-joining analysis：近隣結合解析
neighbor-joining tree：近隣結合系統樹
neighboring population：隣接する個体群
nematode：線虫、ネマトーダ
Neo-Darwinism：ネオダーウィニズム、新ダーウィン説
neo-Hopkins principle：ネオ - ホプキンス則、新ホプキンス則
neolignoid feeding deterrent：ネオリグノイド摂食抑制剤
neonate：新生幼虫
Neoptera：新翅目、新翅群
neopterobilin：ネオプテロビリン
neotenic reproductive：ネオテニック生殖虫、幼形成熟の生殖虫
neoteny：幼形成熟、ネオテニー
Neotropic region：新熱帯区

neotropical nocturnal butterfly：新熱帯産の夜行性蝶
Neotropical region：新熱帯区
neotropics：新熱帯種、新熱帯
neotype：ネオタイプ、新基準標本
nerve branch：神経枝
nerve cord：神経索
nerve ending：神経終末
nerve ganglion：神経球
nerve root：神経根
nerve stain：神経用染料、神経用染色
nervulus：前区脈、肘臀横脈
nest：巣
nesting site：営巣場所
nestmate recognition：巣仲間認識
net：捕虫網
net competitive effect：正味の競争効果
nettle-feeding larva：イラクサ摂食幼虫
network tree：ネットワーク樹
neural activity：神経活動
neural recruitment：神経(的)漸増
neural response：神経応答、神経反応
neural signal：神経信号、神経シグナル
neural spike：神経スパイク
neural stimulus：神経刺激
neurobiological difference：神経生物学的差異
neuroendocrine control：神経内分泌調節
neuroendocrine mechanism：神経内分泌機構
neuroendocrine regulation：神経内分泌制御
neuroendocrine system：神経内分泌系
neuroethology：神経行動学
neurohormonally：神経ホルモン的に
neuromuscular activity：神経筋活動

neuronal connection：神経結合
neuropeptide hormone：神経ペプチドホルモン
neurophysiological recording：神経生理的記録法
Neuroptera：アミメカゲロウ目、脈翅目
neutral evolution：中立(的)進化
neutral insect：ただの虫
neutral site：中立(的)部位
neutralizer：中和物、中和剤
new and revised edition：新訂版
new and unrecorded butterfly：新規の未記載蝶
new combination：新組合せ
new infestation：新たな蔓延
new replacement name：新置換名
new scientific name：新学名
New World tropics：新世界熱帯
newly eclosed virgin female：新しく羽化した未交尾の雌
newly evolved Plantago-feeding population：新しく進化したオオバコ科を摂食する個体群
newly hatched larva：新しく孵化した幼虫
next-less-complex model：一段階下げた複合モデル
next-more-complex model：一段階上げた複合モデル
niche：ニッチ、生態的地位
niche shift：ニッチシフト、ニッチ分化
nicotinic acetylcholine receptor：ニコチン性アセチルコリン受容体
night-flying moth：夜行性の蛾
night-length：夜長時間
NIL：近似同質遺伝子系統
　(Near-Isogenic Line)

nitrate ion：硝酸イオン
nitrogen：窒素
nitrogen requirement：窒素要求量
nitrogen-containing inorganic ion：窒素含有無機イオン
nitrogen-deficient diet：窒素欠乏食料
nitrogen-fixing food plant：窒素固定食草、窒素固定食餌植物
nitrogenous compound：窒素化合物
nitrogenous substance：窒素物質
NMR：核磁気共鳴（Nuclear Magnetic Resonance）
no conflicting linkage relationship：無矛盾の連鎖関係
no significant correlation：無有意相関
no significant difference：無有意差
NOAA：アメリカ海洋大気庁（National Oceanic and Atmospheric Administration）
noctuid moth：ヤガ科の蛾
noctuid moth species：ヤガ科の種
nocturnal：夜行性の、夜間の
nocturnal behavior：夜行活動
nocturnal lifestyle：夜行性の生活様式
nocturnal moth-like ancestor：夜行性蛾のような祖先
nocturnal prosimian：夜行性原猿
nocturnality：夜行性
nocturnally active：夜に活動的な、夜間活動する、夜行性の
nodule：こぶ、瘤
nom.：学名命名（nomen）、名前
nom. nov.：新名（nomen novus）
nomen dubium：（複．nomina dubia）、疑問名〔ラテン語〕
nomen novum：（複．nomina nova）、新名〔ラテン語〕
nomen nudum：（複．nomina nuda）、裸名〔ラテン語〕
nomen oblitum：（複．nomina oblita）、遺失名〔ラテン語〕
nomen protectum：擁護名〔ラテン語〕
nomenclator binominalis：二名法、二名式命名法
nomenclator trinominalis：三名法、三名式命名法
nomenclatural：命名法的
nomenclatural act：命名法的行為
nomenclatural status：命名法的地位
nomenclature：命名法
nominal family-group taxon：名義科階級群タクソン
nominal genus：名義属
nominal taxa：名義タクソン名、名義タクソン
nominate：名義タイプの、承名の
nominate subspecies：名義タイプ亜種
nominotypical taxon：名義タイプタクソン
non-calculated migration：無目標移動
non-consumptive effort：非消費効果
non-diapause condition：不休眠条件
non-diapause destined larval period：非休眠になる場合の幼虫発育期間、非休眠性幼虫発育期間
non-diapause egg：非休眠卵、不休眠卵
non-diapause larva：非休眠幼虫、不休眠幼虫
non-diapause selection：不休眠選択
non-diapausing adult：不休眠成虫
non-diapausing generation：不休眠世代
non-diapausing individual：非休眠個体
non-diapausing pupa：非休眠蛹、不休眠蛹

non-*Drosophila* system：非ショウジョウバエ様式
non-fertile sperm：受精能力をもたない精子
non-flier form：非移動型
non-focal ablation：フォーカス以外の部位切除
non-freezing cold injury：凍傷以外の低温障害
non-innervated microtrich：非神経支配の微毛
non-invasive：侵略性ではない、非侵略性
non-linearity：非線形性
non-migrant：非移動性種
non-migratory butterfly species：非移動性蝶種
non-native species：非在来種
non-overlapping generation of adult：成虫の離散世代
non-periodic form：無周期型
non-poisonous species：無毒種
non-preference：非選好性
non-random dispersal：非ランダム分散、非任意分散
non-socketed hair：ソケットなしの毛、ソケットを付けない毛
non-sperm protein：無精子タンパク質
non-synonymous change：非同義置換
non-trophic interaction：食う食われる以外の関係
non-uniform diapause：不斉休眠
non-uniformity：不斉一化、不斉一性
non-volatile compound：不揮発性化合物
nonadaptive recombinant：非適応的組み換え種
nonasymptotic richness：非漸近的種数
noncoding region：非翻訳領域、非コード領域
nonindigenous species：非在来生物種
nonlinear：非線形の
nonmimetic intermediate：非擬態中間型種
nonmimetic phenotype：非擬態的表現型
nonmimetic recombinant：非擬態型組み換え種
nonnative species：外来種
nonparalogous gene：非パラロガス遺伝子
nonparametric estimator：ノンパラメトリック推定量
nonplastic, dorsal eyespot：背側の非可塑性の眼状紋
nonsound-producing：音を発生しない
nontreated insect line：無処理の昆虫系統
normal activity：平常活動
normal activity range：平常な行動範囲
normal band：正常な縞状バンド列
normal brood：正常な同腹仔
normal emergence rate：正常な羽化率
normal female reproductive organ：正常な雌性生殖器官
normal form：正常型
normal matriline：正常な母系群
normal ovary：正常な卵巣
normal pupa：正常蛹
normal sex ratio：正常な性比
normal testis：正常な精巣
normalization：規格化、正規化
normalized absorbance spectrum：規格化された吸収スペクトル
normalizing selection：正常化選択
normally spread wing：正常に伸びた翅
North American buckeye butterfly：アメリカタテハモドキ

northerly flight：北方へ飛行
northern character：北方的性格
northern continent：北方大陸
northern edge of distribution：分布北限
northern hemisphere：北半球
northern Japanese population：北日本個体群、北日本産個体群
northern margin：北端、北縁部
northern origin：北方起源、北方系
northern range expansion：北への分布拡大
northern species：北方的な種
northernmost area：北限地域、最北端地域
northernmost population：最北個体群、最北端の個体群
northward invasion：北方への侵入
northward spread：北方への広がり
notch：刻み、くぼみ、ノッチ
notched：鉤爪状の
noteworthy：顕著である
noticeable：著しい
notum：背板、中胸背楯板
notwithstanding：それにもかかわらず、にもかかわらず
noun phrase：名詞句
nourishing substance：栄養物質
nov.：新規の（novus）、新しい
nuclear gene：核遺伝子
nuclear magnetic resonance：核磁気共鳴、NMR 解析
nucleate spermatozoon：有核精子
nucleotide datum：（複．-ta）、塩基データ、ヌクレオチドデータ
nucleotide divergence：塩基分岐、ヌクレオチド分岐
nucleotide diversity：塩基多様性、ヌクレオチド多様性
nucleotide mixture solution：ヌクレオチド混合溶液
nucleotide position：塩基位置、ヌクレオチド位置
nucleotide sequence：塩基配列、ヌクレオチド配列
nucleotide sequence database：塩基配列データベース
nucleotide site：塩基部位、ヌクレオチド部位
nucleotide variable position：塩基の位置番号、ヌクレオチドの位置番号
nucleus：（複．-clei）、核、細胞核
nucleus of bursa copulatrix cell：交尾のう細胞の細胞核
nucleus of Malpighian tubule cell：マルピーギ管細胞の細胞核
nudum：裸〔ラテン語〕
null hypothesis of no introgression：無遺伝子移入の帰無仮説
null model：帰無仮説モデル、ヌルモデル、独立モデル
number of adult emerging：羽化した成虫の個数、成虫羽化数
number of base substitution：塩基置換数
number of frost day：氷点下の日数
number of genera：属数
number of generation：発生回数、世代数
number of incoming individual：飛来個体数
number of individual：個体数
number of investigation：調査回数
number of investigation individual：調査個体数
number of larval instar：幼虫の齢数
number of oviposition：産卵数

number of plants/area：植物数／面積
number of rainy day：降水日数
number of silk girdle：帯糸数
number of species：種数
number of stem：樹幹数、幹数、本数
numerical response：数の反応、数量的反応
nuptial gift：婚姻贈呈物
nursery stock：苗木
nutrient：栄養になる食物
nutrient budget：栄養物質の収入分
nutrient cycling：栄養循環
nutrition：栄養
nutritional condition：栄養条件
nutritional ecology：栄養生態学、栄養学的生態
nutritional mechanism：栄養学的機構
nutritional requirement：栄養要求、栄養要求性
nutritional suitability：栄養的適合性
nutritious accessory gland product：栄養に富んだ付属腺物質
nylon membrane：ナイロン膜
nymph：若虫（わかむし）、ニンフ、幼虫
nymphalid branch：タテハチョウ科の分岐

o

oak：オーク（カシ、ナラ、カシワ）類の木
oak tiger butterfly：オビモンドクチョウ
object：対象物
objective：客観的
objective synonym：客観（的）同物異名
obligate association：絶対共生的関係
obligatory：内因性
obligatory diapause：内因性休眠
obligatory migration：内因性移動

OBP：臭物質結合タンパク質（Odorant Binding Protein）
observation：観察
observation-area curve：観察領域の曲線
obstruct：ふさぐ、塞ぐ
obtect：殻で覆われた
obtect pupa：被蛹（ひよう）
obtuse：鈍角
occasional phenotypic leap：偶発的表現型の飛躍
occasional recombinant phenotype：偶発的組み換え表現型
occupancy：占有性
occupant：占有者
occupied area：占有地域
occupied patch：占有パッチ
occupied site：占有地
occupied space：占有空間
occupying individual：占有個体
occur：発生する
occurrence of sexual mosaic：性モザイクの発生
Oceanian region：オセアニア区
ocellar：単眼の
ocellated marking：眼状紋
ocellus：（複 . -li）、単眼、眼状紋〔ラテン語〕
octopamine：オクトパミン
octopamine receptor：オクトパミン受容体
odds of diapause：休眠の確率値、休眠のオッズ
Odonata：トンボ目、蜻蛉目
odor substance：臭物質
odorant binding protein：臭物質結合タンパク質
odoriferous compound：芳香（性）化合物
oenocyte：エノサイト

oenocytoid：エノシトイド（昆虫血球細胞の一種）
offense：罪、違反、攻撃
Official Correction：公式訂正書
official text：正文、公式原文
offprint：別刷り、抜き刷り
offspring：子孫、子
OFT：最適採餌理論（Optimal Foraging Theory）
oily droplet：オイル液滴
old-growth forest stand：老齢林分
olfactory cue：臭覚刺激
olfactory gene：臭覚遺伝子
olfactory receptor：嗅覚感覚器、嗅覚受容体
olfactory sense：嗅覚
olfactory stimulus：臭（覚）刺激
oligopause：弱い休眠、仮性休眠
oligophagous：狭食性、狭食性の、少食
oligophagy：狭食性
ommatidial unit：個眼ユニット
ommatidium：（複．-tidia）、個眼〔英語〕
ommatine D：オマチン D
ommine：オミン
ommochrome：オモクローム
ommochrome pathway gene：オモクローム系の経路遺伝子
ommochrome pigment：オモクローム系色素
omniphagy：雑食性
omnivore：雑食（性）動物
on line：蝶道型
once-abandoned woods：一度見捨てられた森
one brood：一化
one generation a year：年一回発生、年一世代
one-step mutation：一段階突然変異
one-way ANOVA：一元配置分散分析
one-way migration：片道移動
ongoing area of research：発展中の研究分野
ongoing decline：進行している衰亡
onset：開始
onset of diapause：休眠の開始
oocyte presence：卵母細胞の存在
oogenesis：卵形成
oogenesis-flight syndrome：卵形成 - 飛翔症候群、卵形成 - 飛翔形質群
open bar：白色の棒
open documentation：公開文書
open dot：白丸、丸点
open population：開放個体群
open question：未解決な問題
open reading frame：オープンリーディングフレーム、読み取り枠
open space：開けた場所、開放的場所
open sunny habitat：開けた陽のあたる生息地
open vegetation：疎生植生
open woodland：疎林地帯
open-line：開線、二重線
opening：木のまばらな空地
operating principle：作動様式、操作原理
Opinion：意見書
opportunistic diapause：日和見休眠
opportunistic life cycle strategy：日和見的な生活史戦略
opposite of the syndrome, the：逆症候群
opsin allele：オプシン対立遺伝子
opsin expression pattern：オプシン発現様式

opsin gene family：オプシン遺伝子ファミリー
opsin gene sequence：オプシン遺伝子配列
opsin protein：オプシンタンパク質
optical signal：光刺激
optimal foraging：最適採餌
optimal foraging theory：最適採餌理論
optimal patch use time：最適餌場滞在時間
optimal solution：最適解
Or：嗅覚受容体（Olfactory receptor）
oral：口側、口部
orange albatross butterfly：ベニシロチョウ
orange sulphur butterfly：アメリカオオモンキチョウ、オオアメリカモンキチョウ
orange tip butterfly：クモマツマキチョウ
orange-barred giant sulphur butterfly：ベニモンオオキチョウ
orangedog butterfly：クレスフォンテスタスキアゲハ、オオタスキアゲハ
order：目（分類階級の「もく」）
ordinary scale：普通鱗（片）
ordinary year：平年
ordinary zone：普通地域
ordination：配列
Oregon silverspot butterfly：オレゴンギンボシヒョウモン
organella：細胞内器官
organic chemist：有機化学者
organism：生物、生体、有機体
organismic mechanism：生体機構
organization：機構
Oriental and tropical origin：東洋熱帯起源
Oriental region：東洋区
Oriental tropical zone：東洋熱帯地域
orientation：定位、オリエンテーション
origin：原点、起源、原産国

origin of diversity：多様性の起源
origin of Eurasian mountains：ユーラシア山岳起源
origin time：起源年代
original description：原記載
original designation：原指定
original native wildlife：本来の在来野生生物
original publication：原公表
original spelling：原綴(つづ)り
ornithologist：鳥類学者
orogenic movement：造山運動
orphan receptor：オーファン受容体
orthogonal 3-way design：直交三元計画
ortholog：直系遺伝子、真正相同遺伝子
orthologue pair：直系遺伝子対
Orthoptera：バッタ目、直翅目
osmeterial secretion：臭角分泌物
osmeterium：（複．-teria）、臭角〔英語〕
other family：他の科
other species：他の種
out-compete native organism：在来生物との競争に勝ち残る
outbreak：大発生
outbreed laboratory population：研究室の非近交系個体群
outbreeding：外交配、非血縁者との交配
outbreeding depression：外交配弱勢、異系交配弱勢、他殖弱勢
outbreeding species：外交配種、異系交配種
outcropping：露出、発生
outdoor insectary：野外の昆虫飼育場
outer angle：後角部、肛角部、後縁角、肛角
outer margin：外縁
outer membrane：外膜

outer membrane surface：外膜表面
outgroup：外群、アウトグループ
outside air：外気
outspread：翅を開いた
oval-shaped：卵形の、卵円形の
oval-shaped outer membrane：卵形の外膜
ovarian development：卵巣発育
ovarian diapause：卵巣休眠
ovarian dormancy：卵巣休眠
ovarian dynamics：卵巣の動態、卵巣動態
ovarian maturation：卵巣成熟
ovariole：卵巣小管
ovary：卵巣、子房
over-represent：大きな比率を占める
overemphasize：過度に強調する
overgrown：生い茂った
overheat：過熱
overlap spatially：空間的に重ね合う
overshooting：行き過ぎ現象
oversummering：越夏
overt defense：表立った防衛
overwinter：越冬する
overwintered generation adult：越冬世代成虫
overwintering：越冬
overwintering ability：越冬能力
overwintering aggregation：越冬集団
overwintering form：越冬型、越冬形態
overwintering larva：越冬幼虫
overwintering season：越冬期
overwintering stage：越冬期、越冬齢期、越冬態、越冬ステージ
oviduct：輸卵管
oviparous：卵生の、卵を生む
oviparous orifice：産卵孔
oviposit：産卵する

oviposition：産卵
oviposition activity：産卵活性
oviposition behavior：産卵行動
oviposition deterrent：産卵抑制物質
oviposition plant：産卵植物
oviposition preference：産卵選好、産卵選好性
oviposition regulator：産卵調節物質
oviposition repellent：産卵忌避物質
oviposition stimulant：産卵刺激物質
oviposition stimulant reception system：産卵刺激物質受容系
oviposition-deterring pheromone：産卵抑制フェロモン
oviposition-grass preference：産卵草選好性
ovipositional behavior：産卵行動
ovipositor：産卵管
ovipositor sheath：産卵管鞘、産卵鞘
ovule：胚珠
ovum：（複.-va）、卵（らん）、卵細胞、卵子〔ラテン語〕
owl：フクロウ
owl butterfly：イドメネウスフクロウチョウ
oxgen toxicity：酸素毒性、酸素中毒
oxyanion：オキシアニオン
oxygen absorption：酸素吸収
oxygen consumption：酸素消費（量）

p

Pa：パスカル（圧力や応力の単位）（Pascal）
PA：ピロリジジンアルカロイド（Pyrrolizidine Alkaloid）
paddy field：水田地帯
painted lady butterfly：ヒメアカタテハ

painted-on eyespot：絵の具で描かれた眼状紋
pair-clinging behavior：組しがみつき行動
paired t-test：対になったt検定
pairwise coancestry coefficient：対比較コアンセストリー係数、対比較共祖係数
pairwise competitive interaction：一対の競争的相互作用
pairwise genetic distance：対比較した遺伝距離、対様式遺伝的距離
palae-：旧 -、古 -〔ギリシャ語〕
Palaearctic region：旧北区
palaeo-：旧 -、古 -〔ギリシャ語〕
palaeontological site：古生物学の場所
palaeotropical zone：旧熱帯
palaeotropics：旧熱帯
palatability：美味性、嗜好性
palatability spectrum：おいしさのスペクトル、嗜好性の範囲
palatable：美味
pale beige ventral hindwing：腹側の青白いベージュ色の後翅
pale forewing：青白い前翅
pale form：青白い型
pale medial band：中央部の青白い縞状バンド
pale summer form：青白い夏型
pale-：旧 -、古 -〔ギリシャ語〕
pale-yellow：淡い黄色
Palearctic region：旧北区
paleo-：旧 -、古 -〔ギリシャ語〕
paleobotanical：古植物学の
Paleogene period：古第三紀
paleovegetation reconstruction：古植生の再構築
palmar：掌側（しょうそく）の、手掌の

palp：鬚（ひげ、しゅ）〔英語〕
palpus：（複 .-pi）、鬚（ひげ、しゅ）〔ラテン語〕
palynological：花粉学の
palynophagy：花粉食性
PAML：最大節約法に基づいて祖先型推定を行うソフトウェアの名称
pan of water：水を張った皿、水のパン
pan trap：パントラップ、平皿トラップ
Pan-Japan Sea Area：周日本海地域
Pan-tropical type：汎熱帯型
Panamanian forest：パナマの森
panmictic：任意配配
panmictic population：汎生殖個体群、任意交配集団
panmictic unit：任意交配単位
paper board：台紙
paper kite butterfly：オオゴマダラ
paper model：色紙モデル
papiliochrome：パピリオクローム
papilla analis：肛乳頭、肛乳房突起
pappus：冠毛
pappus length of achene：痩果の冠毛の長さ
para-：側 -〔ギリシャ語〕
parabiosis：併体結合、パラビオーシス
paradigm shift：パラダイムシフト、範例シフト
paradoxical：逆説的な
parafilm：パラフィルム
paralectotype：パラレクトタイプ、副後基準標本
parallel amino acid change：アミノ酸の平行的変化
parallel change analysis：平行的変化解析
parallel cladogenesis：平行分岐進化
parallel evolution：平行進化、並行進化

parallel phenotype：平行表現型、同位表現型

parallel phenotypic evolution：平行的表現型進化、表現型の平行進化

parallel radiation：平行発散

parallelism：平行進化

paranodal plate：前胸板

parapatagium：側頸板

parapatric：側所的

parapatric distribution：側所的分布

parapatric race：側所的亜種、側所的系統

paraphyletic：側系統の

parasite：寄生、寄生虫、寄生者

parasite datum：（複．-ta）、寄生データ

parasitic：寄生的、寄生型

parasitic interaction：寄生相互作用

parasitic thrips：寄生性のアザミウマ

parasitic wasp：寄生蜂

parasitism：寄生関係、寄生

parasitization：寄生

parasitoid：捕食寄生者、捕食寄生虫、捕食寄生（の）

paratype：パラタイプ、副模式標本、従基準標本

parental population：親の個体群、親種

paris peacock butterfly：ルリモンアゲハ

Parnassians：ウスバアゲハ亜科

Parnassians and Swallowtails：アゲハチョウ科

pars intercerebralis：脳間部

parsimonious explanation：控えめな説明

parsimonious tree：節約な系統樹

parsimony：節減、倹約

parsonsia：ホウライカガミ属

parsonsieae：ホウライカガミ族

part：器官、部分

Part of the List of Available Names in Zoology：動物学における適格名リストの分冊

part of wing membrane：翅の膜部分

parthenogenesis：単為生殖（たんいせいしょく）

parthenogenetic induction：単為生殖誘導

partial bleaching of rhabdom：感桿の部分退色、感桿の部分脱色

partial L opsin gene sequence：L オプシン遺伝子の部分塩基配列

partial voltine：部分化性

partially protected species：部分的保護種

partially purified prothoracicotropic hormone：部分精製された前胸腺刺激ホルモン

partivoltine：部分化性

passing female：通過雌

passion-vine butterfly：ヒョウモンドクチョウ

past report：従来の報告

patagium：（複．-ia）、頸板〔ラテン語〕

patch：大斑点、パッチ、生息場所

patch colonization：パッチへの移住

patch network：パッチネットワーク

patch occupation rate：パッチ占有率

patchiness：パッチネス、不調和、むらがある

patchiness parameter：不均一パラメータ

patchy distribution：パッチ状分布

paternal chromosome：父系染色体

paternal investment：雄側の投資

paternity：父性

paternity analysis：父系分析、父性分析

pathway：経路

patrol：占有する

patrolling：巡回型、巡回

patrolling behavior：巡回行動
patrolling flight：巡回飛翔
patrolling type：探索型
pattern：様相、斑紋
pattern formation mechanism：パターン形成機構
pattern of seasonal change：季節変化パターン
pattern variation：パターン変化
Patterson's D-statistic：パターソンのD統計量
pb gene：プロボースィペディア遺伝子（*proboscipedia*；口器足）、吻足遺伝子
PBAN：フェロモン生合成活性化神経ペプチド（Pheromone Biosynthesis Activating Neuropeptide）
PCMH：蛹クチクラ黒色化ホルモン（Pupal-Cuticle Melanizing Hormone）
PCR：ポリメラーゼ連鎖反応、PCR法（Polymerase Chain Reaction）
PCR amplicon：PCRアンプリコン
PCR mixture：PCR混合物
PCR product：PCR産物
PCR temperature profile：PCRの温度プロファイル
PCR-derived probe：PCR由来プローブ
PD axis：近位-遠位軸、PD軸、基部-末端軸、基部-外縁軸（Proximal-Distal axis）
pdb：タンパク質構造データバンク（Protein Data Bank）
peacock butterfly：クジャクチョウ
peak absorbance：ピーク吸収
peak direction：ピーク方向
peak sensitivity：ピーク感度

pear-shaped：セイヨウナシ形の
pear-shaped eyespot：セイヨウナシ状の眼状紋
peat：ピート、泥炭
pecky rice：斑点米
pedicel：梗節（こうせつ）
peer review：ピアレビュー、同等者による査読、同輩審査
pellet：小球を投げつける、ペリット、糞球
penalty：罰則
penetrance：浸透率
penetrated light：透過光
penis：陰茎〔ラテン語〕
Pennsylvanian period：ペンシルベニア紀（地質時代の古生代／石炭紀（Carboniferous period）の一時代）
penultimate instar：亜終齢
peppered moth：オオシモフリエダシャク
peptidoglycan recognition protein：ペプチドグリカン認識タンパク質
PER：口吻伸展反射（反応）（Proboscis Extension Reflex (Response)）
per capita coefficient：1個体当りの係数
per capita interaction：1個体当りの相互作用
per copy of the gene：遺伝子1コピー当り
per-：通して-、非常に-、完全に-〔ラテン語〕
perch site：静止場所
perching：待ち伏せ型、みはり、見張り
perching behavior：静止行動
perching male：静止雄
perching time：静止時間
performance：成育達成度、遂行
performance of estimator：推定量の性能
peri-：周辺-、周囲-〔ギリシャ語〕
peril of normalizing richness：種数を正

規化することの危険
perilous：危険に満ちた
period：紀（地質時代の「き」）
period of irradiation：照射期間
periodic form：周期型
periodic rediscovery：周期的な再発見
periodic response：周期反応
peripatric speciation：周辺（的）種分化
peripheral projection pattern：周辺投射パターン、周辺投影様式
perish：消滅する、消え去る
peritracheal gland：気管周腺
peritrophic membrane：囲食膜（いしょくまく）
permanent diapause：永久休眠
permanent quadrat：永久コドラート、永久方形区
permeate：充満させる、しみ込む
Permian period：ペルム紀（地質時代の古生代の一時代）
persistence：回復力、存続性、持続性
persistence ability：存続能力、回復力
persistent hybridization：存続する交雑、持続する交雑
persisting metapopulation：存続するメタ個体群
personal communication：私信
personality：個性
persuasive evidence：説得力のある証拠
pervasive：広がる、まん延する
pest：害虫、有害な虫
pest control：害虫防除
Petri dish：ペトリ皿、シャーレ
petroleum jelly：ペトロラタム（石油から採る半固形状のろう）、ワセリン
PGC：始原生殖細胞（Primordial Germ Cell）

PGRP：ペプチドグリカン認識タンパク質（PeptidoGlycan Recognition Protein）
phagocytosis：食細胞活動、食作用
phagostimulant：摂食刺激物質
phallus (ph)：挿入器、ファルス
pharate adult：潜成虫
pharmacophagy：薬物食性、薬物摂食
pharynx：咽頭（いんとう）〔ラテン語〕
phase：相、位相
phase adjustment：位相調節
phase gregaria：群生相
phase polymorphism：相多型、相変異
phase related polymorphism：相関連多型、相関連変異
phase solitaria：孤独相
phase transien：転移相
phase variation：相変異
phasic/tonic nature：一過性／持続性の性質
Phasmatodea：ナナフシ目、竹節虫目
phellamurin：フェラムリン
phenanthroindolizidine alkaloid：フェナントリジンアルカロイド
pheno-：表現 -〔ギリシャ語〕
phenol-chloroform：フェノールクロロホルム
phenol-chloroform extraction procedure：フェノール - クロロホルム抽出法
phenology：季節的消長、生物気象学、生物季節学
phenomenon：現象
phenotype：表現型、表現形
phenotype mechanism：表現機構
phenotype variation：表現型変異
phenotypic appearance：表現型発現

phenotypic change：表現型の転換
phenotypic character：表現形質
phenotypic differentiation：表現型分化
phenotypic divergence：表現型分岐、表現型発散
phenotypic diversity：表現型多様性
phenotypic effect：表現型効果、表現型の効果
phenotypic evolution：表現型進化
phenotypic intermediate：中間表現型
phenotypic plasticity：表現型可塑性、表現型可塑性、表現型可変性
phenotypic resemblance：表現型酷似
phenotypic variation：表現型変異
pheromone：フェロモン
pheromone biosynthesis activating neuropeptide：フェロモン生合成活性化神経ペプチド
pheromone precursor：フェロモン前駆体、フェロモン前駆物質
phleophagy：樹皮食性
phorcabilin：フォルカビリン
phospholipid：リン脂質
photo-：光 -〔ギリシャ語〕
photo-receptor：光感覚器、光受容器
photograph：写真を撮る、撮影する
photographic plate：写真図版、写真乾板
photography：写真撮影
photoisomerize：光異性化する
photoisomerizing flash：光異性化閃光
photoperiod：光周期、日長
photoperiod treatment：光周処理
photoperiod-sensitive stage for diapause induction：休眠誘起の日長感受期
photoperiodic clock：光周時計

photoperiodic control：光周調節、日長調節
photoperiodic counter：光周（期）カウンター、光周計数機構
photoperiodic response：光周反応
photoperiodic response curve：光周反応曲線
photoperiodic signal：光周信号
photoperiodic time measurement：光周測時
photoperiodic time-measurement system：光周測時機構
photoperiodically controlled diapause：光周性で調節された休眠
photoperiodism：光周性（こうしゅうせい）、日長効果
photophase：明期
photophil：親明相
photopigment：視物質、感光色素、光色素
photopigment sensitivity：視物質感度
photopigment-containing microvillar membrane：光色素含有微絨毛膜
photoreceptive interneuron：光感受性介在神経
photoreceptor：光感覚器、光受容器
photoreceptor cell：光受容細胞
photoreceptor nucleus：光受容細胞核
photosensitive part：感光部位
photosensitive portion：感光部分
photosynthesis：光合成
phototaxis：走光性（そうこうせい）
photothermograph：光温図表
phyllophagy：葉食性
phylogenetic analysis：系統解析
phylogenetic biogeography：系統生物地理学

phylogenetic biogeography method：系統生物地理学的方法
phylogenetic branch：系統分岐
phylogenetic method in biogeography：系統生物地理学的方法
phylogenetic reconstruction：系統再構築、系統再構成
phylogenetic relationship：系統（的）相互関係、系統関係
phylogenetic scale：系統学的な尺度
phylogenetic study：系統分類に関する研究、系統学的研究
phylogenetic tree：系統樹
phylogenetic utility：系統学的効用
phylogenetic view point：系統的視点
phylogenic evolution：系統進化
phylogeny：系統発生（進化）、系統分類、系統学、系統
phylogeny reconstruction：系統発生の再構築、系統発生の再構成
phylogeographic history：系統地理学的歴史
phylogeographic interpretation：系統地理学的解釈
phylogeography：系統地理学、系統地理
phylogram：系統図、系統樹
phylum：（複．-la）、門（分類階級の「もん」）
PHYML：最尤法（maximum likelihood）のソフトウェアの名称
physical characteristic：物理的性格
physical control：物理的防除
physical difference：物理的相違
physical factor：物理的要因
physical linkage：物理的連鎖
physical-to-map distance：物理的距離

physiological adaptation：生理（学）的適応
physiological and behavioral experiment：生理学的・行動学的実験
physiological approach：生理学的アプローチ
physiological background：生理的背景
physiological characteristic：生理的特性、生理特性
physiological cost：生理的コスト
physiological datum：（複．-ta）、生理データ
physiological mechanism：生理機構
physiological mediation：生理的媒介
physiological recording：生理（学）的記録
physiological regulation：生理的調節
physiological response：生理的反応
physiological resurgence：生理的誘導多発生
physiological saline：生理食塩水
physiological salt solution：生理食塩水
physiological state：生理状態
physiological study：生理学的研究
physiological time：生理（学）的時間
physiological trait：生理的形質、生理的特性
physiologically regulated step by step：生理的に調整された一連の行動
physiology：生理学、生理
phyto-：植物-、植物の-〔ギリシャ語〕
phytoalexin：ファイトアレキシン
phytochemical concentration：植物化学物質濃度
phytoecdysone：植物エクジソン
phytophagous myrmecophile：植食性の好蟻性昆虫
phytophagy：植食性

pierid butterfly：シロチョウ科の蝶
pierisin：ピエリシン
pigment：色素
pigment content：含有色素
pigment deposition：色素沈着
pigment fraction：色素分画
pigment layer：色素層
pigment pathway：色素（生合成）経路
pigment polymorphism：体色多型、色素多型
pile：山積み
pinaculum：（複．-la)、硬皮板、有棘毛瘤〔ラテン語〕
pine beauty moth：マツキリガ
pine white butterfly：マツノキシロチョウ
pinewood：松林
pioneer：先駆種、開拓者
pioneer individual：パイオニア個体、先駆個体
pioneer plant：先駆樹種、先駆植物
pipevine：ウマノスズクサ
pipevine swallowtail butterfly：アオジャコウアゲハ
pitfall：ピットフォール、落し穴、注意点
pivotal experimental datum：（複．-ta)、極めて重要な実験データ
place of occurrence：発生地
place of origin：発祥地
placement of sample：サンプルの配置
placental mammal：有胎盤哺乳類
plant as oviposition site：産卵植物
plant bug：カスミカメムシ、カメムシ
plant community：植物群集、植物群落
plant density：植物密度
plant ecologist：植物生態学者
plant nursery：園芸店

plant odor：植物香気、植物の香り
plant outlier：植物の異常値
plant pigment：植物色素
plant protection：植物保護
plant quarantine：植物検疫
plant species：植物種
plant trichome：植物の毛状突起
plant worm：冬虫夏草
planter：底側(ていそく)
planthopper：ウンカの仲間
planting：植林
plasmatocyte：プラズマ細胞（血球細胞の一種）
plasmid：プラスミド
plasmid purification method：プラスミド精製法
plastic cup：プラスチックカップ容器
plasticity：可塑性
plausible explanation：もっともらしい説明
playback：録音の再生
Plecoptera：カワゲラ目、襀(せき)翅目
pledge：誓約する
pleiotropic effect：多面発現効果
pleiotropy：多面発現
Pleistocene climate change：更新世の気候変動
Pleistocene refugium：更新（洪積）世での隔離
plenary power：強権
plesiomorph：原始的初原形質、原始的旧形質
pleurite：側板
pleuron：側板(そくばん)
ploidy：倍数性、倍数関係
ploidy level：倍数性レベル

plume：冠毛

pluvial：多雨の

PMRF：蛹黒色化抑制因子（Pupal Melanization Reducing Factor）

PNP：前胸板（ParaNodal Plate）

poikilotherm：変温動物

point of adult emergence：成虫羽化の時点

pointed apex in forewing：前翅頂のとがり

pointed tip：とがった先端

poison：毒

poison-antidote model：毒性 - 解毒モデル、毒 - 毒消しモデル

poisonous：有毒な

polar coordinate model：極座標モデル

polarisation-dependent color vision：偏波依存色覚

polarized：分極化された、極性を示す

polarized light：偏光

poleward shift：極方向への移動

pollen：花粉〔ラテン語〕

pollen dispersal：花粉飛散、花粉分散

pollen feeding：花粉摂食

pollen flow：花粉流動

pollen resource：花粉資源

pollination：受粉、花粉媒介

pollination ecology：受粉生態学、送粉生態学

pollination syndrome：送粉シンドローム

pollinator：花粉媒介者、花粉媒介昆虫、ポリネーター、送粉者

pollinator-friendly species：花粉媒介者が好む種

pollinophagy：花粉食性

pollution：汚染

poly-：多 -〔ギリシャ語〕

polyandry：一雌多雄、一妻多夫

polyembryony：多胚形成

polygene：多遺伝子、量的遺伝子、ポリジーン

polygene control：ポリジーン支配

polygenic：多遺伝子（性）の、ポリジーンの

polygyny：一夫多妻、ポリガミー、複女王性

polyhedrosis：多角体病

polymerase chain reaction：ポリメラーゼ連鎖反応（PCR）

polymorphic：多形型、多型、多型の、多型的

polymorphic Batesian mimic：多型のベイツ型擬態種

polymorphic population：多型個体群

polymorphic race：多型系統

polymorphic site：多型部位

polymorphism：遺伝的多型、多型、多型現象、多型性

polypeptide factor：ポリペプチド因子

polyphagous：広食性の、広食性、多食性

polyphagy：広食性

polyphenism：表現型多型、環境的多型現象

polyphenism phenomenon：多型現象

polyphenol oxidase：フェノールオキシダーゼ

polyphyletic：多系統的、多系統

polyspermy：多精

polytypic：多型の、多型的

polyvoltine：多化性

pomona form：銀紋型（ウスキシロチョウ）、ポモナ型

pond：池

pool：水たまり

pooled individual：プールされた個体
pooled quadrat：プールされた方形区
poor nutrition：栄養不足、粗末な栄養
poor summer weather：悪天候の夏
populate：住む
population：個体群(【生態学】)、集団、個体数、母集団(【統計学】)
population biology：個体群生物学
population bottleneck：集団ボトルネック、集団遺伝学におけるボトルネック効果
population change：個体数の変化、個体群変動
population density：個体群密度
population dynamics：個体群動態
population dynamics model：個体群動態モデル
population ecology：個体群生態学
population genetic analysis：個体群遺伝解析
population genetic approach：個体群の遺伝学的研究法
population genetic structure：個体群の遺伝的構造、集団の遺伝的構造
population growth：個体群成長
population growth rate：個体群成長率、個体数増加率
population isolation：個体群の孤立化
population management：個体群管理
population regulation：個体群調整、個体数調整、個体群制御
population size：個体群サイズ、集団の大きさ
population stability：個体群の安定性
population structure：個体群構造
population survival：個体群の生存
population-genetic approach：個体群遺伝的手法
population-level change：個体群レベルの変動、個体数レベルの変化
populations of individual：個体の母集団
pore cupola organ：有孔の鐘状感覚器
porrect：前に突き出た
port：心門
positional cloning of locus：遺伝子座の位置クローニング、遺伝子座の位置的単離
positional genetic homology：遺伝子の位置相同性
positional information：位置情報
positional value：位置価(分子的番地表示)
positive environmental factor：関与する環境要因、正の環境要因
positive feedback：正のフィードバック
positive selection：正の自然選択
positive selection inference：正の自然選択の継承
positive slope：右上がり斜線、上り傾斜、正の傾斜
possess：持つ、所有する
possession：所有
possibility of copulation at the emergence site of female：雌の羽化場所での交尾可能性
possible overwintering area：越冬可能地域
post teneral flight：テネラル(羽化)以後の移動飛翔
post teneral period：羽化以後期
post-：後-、外-〔ラテン語〕
post-border：国境通過後
post-diapause larva：休眠後幼虫
post-eclosion：羽化後

post-mating isolation：交配後隔離
postbasal：外基部帯
postdiscal：外中央帯、外縁寄り中央帯
postdiscal band：外中央帯
postdiscal line：外横線
posterior：後、後部の、後方の〔ラテン語〕
posterior apophyses：後側甲
posterior branch：後枝
posterior crossvein：後横脈
posterior cubitus vein：中脈分岐、CuP脈
posterior edge：後縁、後端
posterior eyespot：後方の眼状紋
posterior focus：後方のフォーカス
posterior side：後側
posteriorly：後方に
postman butterfly：ベニモンドクチョウ、メルポメーネベニモンドクチョウ
postman form：ポストマン型
postmating isolation：交配後隔離
postmedial：外中央帯、翅外中央帯
postmedian：外中央部
postnatal seta：出生後の刺毛
postreceptoral：後受容の
postulate：仮定する
postural change：姿勢変更
posture：姿勢
postzygotic isolation：接合後隔離
pot：鉢
potassium hydroxide：苛性カリ、水酸化カリウム
potent source：有力な源泉
potential：潜在力、可能性
potential candidate：潜在的な候補（遺伝子）
potential hearing organ：潜在的な聴覚器官

potential high-risk species：潜在的にリスクの高い種
potential insect pest：潜在（的）害虫
potential of hydrogen：pH、ペーハー
potential reproductive success：潜在繁殖成功度
potential threat：潜在的脅威
potentially invasive species：潜在的侵入種
potentially valid name：潜在的有効名
poverty：貧困
power of ANCOVA：共分散分析の検定力、共分散分析の検出力
powerful flier：力強く飛ぶ
pre-：前-〔ラテン語〕
pre-adult stage：幼若成虫期、前成虫期
pre-border：国境通過前
pre-chilling incubation period：前低温馴化期間
pre-epimeron：前方後側板
pre-freezing：予備凍結
preadaptation：前適応
preadaptation to host plant：寄主植物に対する前適応
preadaptation to survive winter in the temperate climate：越冬を可能にする気候前適応
precautionary approach：予防的アプローチ、予防的手法
precedence：優先権
precipice：崖
precipitation：降水量
precocious pupa：早熟蛹
precostal vein：肩脈
predation：捕食
predation pressure：捕食圧
predator：捕食（性）動物、肉食（性）動

物、捕食者、捕食虫
predator avoidance：捕食者逃避
predator avoidance behavior：対捕食(者)回避行動
predator evasion：捕食者回避
predator-avoidance：捕食者回避
predator-avoidance response：捕食者回避反応
predatory bird：捕食(性)鳥
prediapause copulation：前休眠期の交尾、休眠前の交尾
predictable：予測的、予測可能な
predictable environment：予測可能な環境
prediction：予測
predictor：予測因子、予測変数
predominant signal：支配的なシグナル、支配的信号
predominantly：支配的で、有力で
predominate：圧倒的に多い
preference：選り好み、選好
preference cue：選好刺激
preference index：選好性指標、選好指標、選好指数、選択指数
preference intermediate to the parental species：親種の中間種を好む
preference locus：選好に関する遺伝子座
preferentially：優先的に、選択的に
prefix：接頭語
preimaginal epidermal organ：幼虫期の表皮器官
preliminary study：事前研究、予備的研究
premating isolation：交配前隔離
preovipositional period：産卵前期
preparation：標本(実験・解剖用動物の)、プレパラート、調製、準備行動
prepattern：予備パターン、プレパターン

preprint：前刷り
prepupa：(複.-ae)、前蛹(ぜんよう)〔ラテン語〕
prepupal：前蛹の、前蛹状態
presence：存在
presence or absence of a Vogel's organ：フォーゲル器官の存否
pressure delivery system：圧力運搬手段
presumptive locus：原基遺伝子座
pretarsus：前跗節(ぜんふせつ)
prevailing method：広く行われている方法
prevailing usage：慣用法
prevalent：流行している
prevention：予防
previous paper：前報
previously-threatened butterfly：以前に絶滅の恐れがあった蝶
prey：えじき、獲物、被食者、捕食する
prezygotic isolation：接合前隔離
primary attractant：一次誘引物質
primary consumer：一次消費者
primary culture：初代培養
primary homonym：一次(異物)同名
primary metabolite：一次代謝物、一次生産物
primary production：一次生産
primary reason：一次的な理由、主要な理由
primary sensory neuron：一次感覚神経、一次感覚ニューロン
primary seta：一次刺毛
primary wing nerve branch：翅の一次神経枝
primate：霊長類
primate color vision system：霊長類の色覚系

primate lineage：霊長類の系統
primate pigment：霊長類の色素
prime purpose：主要な目的
primer：プライマー
primer pheromone：起動フェロモン、プライマーフェロモン
primitive：原始的な
primitive form：原始的な形態
primitive species：原始的な種、原始種
primordial germ cell：始原生殖細胞
principal component analysis：主成分分析
principal factor：主要因
principle：原理
Principle of Binominal Nomenclature：二名法の原理
principle of competitive exclusion：競争排除則
principle of coordination：同位の原理
principle of homonymy：同名関係の原理
principle of priority：先取権の法則、先取権の原理
principle of the first reviser：第一校訂者の原理
principle of typification：タイプ化の原理
printer's error：印刷者の過誤
printing on paper：紙への印刷
prior permission：事前の許可
priority：プライオリティ、先取権
priority of a name：学名の先取権
priority of a nomenclatural act：命名法的行為の先取権
pro-：前 -〔ギリシャ語〕
probable model：可能性のあるモデル
probe cleanup：プローブクリーンアップ
proboscis：口吻（こうふん）〔ラテン語〕
proboscis extension reflex response：口吻伸展反射反応、口吻を伸ばしての食物探索
proboscis recoil：口吻収縮
process：プロセス、過程、突起、隆起
processes of migration, extinction, and recolonization：移動・絶滅・再定着プロセス
procurement：獲得
producer：生産者
production of all-male brood：すべて雄の同腹仔の産出
productivity gradient：生産性勾配
profound scholar：碩学（造詣が深い博学者）
profoundly：おおいに、深遠に
progeny：子世代、子孫、後代〔ラテン語〕
programmable attenuator：プログラム（化）できる減衰器
programmed cell dead：プログラムされた細胞死
progression of diapause：休眠の進行
project laterally：側方に投影する
projection：つの、突出（突起）部
proleg：（幼虫の）腹脚（ふくきゃく）、前脚（ぜんきゃく）
proleucocyte：原白血球（昆虫血球細胞の一種）
proliferation pattern：増殖パターン
proline-rich：高プロリン含有
prolongation：延長、延長部分
prolongation of the corner of the wing：翅の辺縁部の延長部
prolonged copulation：長期間の交尾
prolonged preovipositional period：長期の産卵前期間
prolonged selection：延長選択

prominent：浮き出ている

prominent dome-shaped inner membrane：突起したドーム形の内膜

prominent eyespot：目立つ眼状紋

prominent projection：突起部

prominent veining：隆起した翅脈

prominent wing vein：盛り上がった翅脈

promiscuous exchange：無差別な交換

promising avenue：前途有望な研究分野

prone to extinction：絶滅する傾向

pronotum：前胸背板(ぜんきょうはいばん)

pronounced change：顕著な変化

pronounced color preference：著しい色彩選好、明確な色彩選好

pronymph：前若虫

propagate outward：外側方向に大きくなる、外側方向に増殖する

propagule：繁殖体、散布体、無性芽

propensity：傾向

propodeum：前伸腹節

proportion of empty patch：空きパッチの割合

proposal：提案書、提唱

prorsa form：プロルサ型(アカマダラ)

prosperity stage：繁栄の段階

protandry：雄性先熟

protected area：保護地域、保護区

protected species：保護種

protective color polymorphism：保護色の色彩多型

protective coloration：保護色

protective coloration effect：保護色効果

protective envelope：かたい殻、保護外被

protective scale：保護鱗粉

protein binding：タンパク質結合

protein biosynthesis：タンパク質生合成

protein body：タンパク質顆粒

protein crosslinker：タンパク質架橋

protein sequence：タンパク質のアミノ酸配列

protein sequence database：タンパク質配列データベース

protein-coding gene：タンパク質コード遺伝子

protein-rich food plant：タンパク質の豊富な食餌植物

prothoracic gland：前胸腺(ぜんきょうせん)

prothoracicotropic hormone：前胸腺刺激ホルモン(PTTH)

prothorax：前胸(ぜんきょう)

protist：原生生物

protozoon：(複. -zoa)、原生動物

protrude：突き出る

provision：条項

provocative：刺激的な

provoke：引き起こす

proximal：近位、基部に近い、中央に近い、基部

proximal ablation：基部側の切除

proximal band：基部側の縞状バンド(内横線)

proximal cause：近因、最も近い原因

proximal region：基部領域、近位領域

proximal-distal axis：近位-遠位軸、PD軸、基部-末端軸、基部-外縁軸

proximate factor：至近要因

proximodistal axis：近位-遠位軸、PD軸、基部-末端軸、基部-外縁軸

proximodistal direction：基部-外縁方向

proximodistally periodic pleat：基部-外縁方向の蛇腹形式に折り畳んだひだ

proxy：代用品

pseudo- : 偽 -、仮 -、擬 - 〔ギリシャ語〕
pseudopupil : 偽瞳孔
Psocodea : カジリムシ目、咀顎目（従来はチャタテムシ目、シラミ目、ハジラミ目に細分されていた。）
psoralen : ソラレン、ソラーレン
pteridine pigment : プテリジン系色素
pteridine ring : プテリジン環
pterin : プテリン系色素
ptero- : 翅 - 〔ギリシャ語〕
pterothoracic ganglion : 翅胸部神経球、翅胸部神経節
PTSH : 前胸腺抑制ホルモン（ProThoracicoStatic Hormone）
PTTH : 前胸腺刺激ホルモン（ProThoracicoTropic Hormone）
public awareness : 普及啓発
public-domain software : パブリックドメインソフトウェア
publication : 公表
publish : 公表する、出版する
published work : 公表された著作物
puddle : 水たまり
puddle butterfly : 水たまりの蝶
puddler : 吸水動物
puddling : 吸水活動
puddling behavior : 吸水行動
pulvillus : （複 .-li)、褥盤（じょくばん）、爪間盤〔英語〕
pumping : ポンピング
pupa : （複 .-ae)、蛹（さなぎ）〔ラテン語〕
pupa period : 蛹期
pupa phase : 蛹期
pupa stage : 蛹期
pupa-diapausing species : 蛹休眠種
pupal coloration : 蛹の色彩化
pupal diapause : 蛹休眠
pupal epidermal sheet : 蛹の表皮層、蛹の表皮シート
pupal mating : 蛹態配偶行動
pupal melanization reducing factor : 蛹黒色化抑制因子
pupal mortality : 蛹死亡率
pupal overwintering : 蛹越冬
pupal period : 蛹期間
pupal weight : 蛹重、蛹の重さ
pupal-cuticle melanizing hormone : 蛹クチクラ黒色化ホルモン
pupal-mate : 蛹態配偶行動
puparium : 蛹殻（ようかく）、囲蛹殻
pupate : 蛹になる、蛹化（ようか）する
pupated individual : 蛹化した個体
pupation : 蛹化（ようか）
pupation rate : 蛹化率
pupation site : 蛹化場所
pupation timing : 蛹化時期
pure brood : 純粋同腹仔
pure cross : 純粋交配種
pure female : 純粋な雌、非雑種の雌
pure form : 純粋型
pure-species : 純粋種
purine derivative : プリン誘導体
purine ring : プリン環
purple emperor butterfly : チョウセンコムラサキ
pursue : 追い求める
pursuit : 追跡
push-pull strategy : プッシュ・プル法
putative binding protein : 推定結合タンパク質
putative exon : 推定上のエクソン、推定されるエクソン

putative photoreceptor organelle：光受容器の可能性をもつ小器官
putative transcription factor：推定転写制御因子
putatively distinct：離れているらしい
putatively relate：関係しているらしい
pyralid moth：メイガ科の蛾
pyramiding：ピラミッディング
pyrrolizidine alkaloid：ピロリジジンアルカロイド

q

QTL：量的形質遺伝子座、量的形質座位（Quantitative Trait Locus）
quadrat：コドラート、方形区、方形枠
quadrat method：コドラート法、区画法、方形区法
quadrat-based richness：方形区数に基づく種数
quadrate wing：角張った翅
quadruped：四足類、四肢類
qualitative substance：質的物質
quality filtering：クオリティーフィルタリング
quantal response：計数的反応
quantification：定量化
quantify：定量化する
quantitative PCR：定量的 PCR 法、定量 PCR
quantitative PCR analysis：定量的 PCR 解析
quantitative prediction：定量的予測
quantitative substance：量的物質
quantitative trait locus：量的形質遺伝子座、量的形質座位
quantitatively inherited：定量的に継承された
quantity：多寡、多量、量、定量
quarantine：検疫
quarantine measure：検疫措置
quarantine regulation：検疫規則
quarantine zone：隔離地域
quarry：採石場、石切場、源泉、種本（たねほん）
quasi-national park：国定公園
Queen Alexandra's birdwing butterfly：アレキサンドラトリバネアゲハ
queen mandibular gland pheromone：ミツバチ女王蜂の大顎腺フェロモン
Queen of Spain fritillary butterfly：スペインヒョウモン
quiescence：静止、休止、活動停止
quinone：キノン系色素

r

R：径脈（Radial vein）
r-selection：r- 淘汰、r 選択
r-species：r 種（r 戦略をとる種）
r-strategy species：r 戦略種
Rab geranylgeranyl transferase gene：Rab ゲラニルゲラニル転移酵素遺伝子
rabble：群れ
race：系統、レース、品種、人種、亜種
race formation：レース形成、系統形成
racial diversification：系統の多様性
racial evolution：系統を生む進化
RAD：制限酵素認識部位に関わる DNA（Restriction-site Associated DNA）
RAD analysis：RAD 解析
RAD resequencing：RAD 再配列決定
radial：径脈の、放射状の
radial crossvein：径横脈、r 脈

radial sector：径分脈、Rs 脈
radial vein：径脈(けいみゃく)、R 脈
radial-medial crossvein：径中横脈、r-m 脈、前横脈
radiant heat：輻射熱
radiate：照射する、放散する
radiation：放射線、放散
radiation of metazoan：後生動物の放散
radius：径分脈、径脈、Rs 脈〔ラテン語〕
radius increase：半径の増分
rainforest：雨林、降雨林
rainy：雨の多い
rainy season：雨期
raise：飼育する、育てる
raising awareness：意識啓発
Rajah Brooke's birdwing butterfly：アカエリトリバネアゲハ
ramped response：ランプ応答
Ramsar Convention：ラムサール条約
random drift：遺伝的浮動、機会的浮動
random mating：ランダム交配、任意交配
random order：ランダム順、無作為順序
random ordering：ランダム順、順不同
random placement curve：ランダム配置曲線
randomized sample accumulation curve：ランダムなサンプルの累積曲線
randomly captured individual：ランダムに捕獲した個体
range：生息域、範囲
range contraction：生息域の縮小
range expansion：生息域の拡大
range fragmentation：範囲の断片化、範囲の分断化
range of latency：潜時帯
range of variation：変異の幅、変異幅

rank：階級、順位
rapid desiccation：急速乾燥
rapid flight：素早く飛び回る、急速飛翔
rapid membrane vibration：膜の高速振動
rapid-assessment survey：迅速評価調査
rapidly radiating clade：急速に放散している単系統
rare：希少
rare hybrid：希少雑種、希少交配種
rare species：稀種、希少種
rare taxa：希少な分類群
rarefaction：希薄化
rarefaction curve：希薄化曲線、希薄曲線
rarefaction formula：希薄化の計算式
rarefaction method：希薄化法
rarefied：希薄な
rash：発疹、かぶれること
rate of migration：移動比、移動率
ratio of richness to area：種数対面積の比
ratio of species/individual：種数／個体数の比
ratio of successful emergence：羽化に成功した割合、羽化の成功割合
rattling noise：カラカラ音
rayed form：放射状型
RBSDV：イネ黒すじ萎縮ウイルス、イネ黒すじ萎縮病（Rice Black Streaked Dwarf Virus）
re-：再 -〔ラテン語〕
re-analysis：再解析、再分析
re-colonise：再移住する、再コロニー化する
re-enforcement：再増強
re-establishment：再定着
re-introduction：再導入
re-sampling：再サンプリング、リサンプ

リング、非復元抽出
re-scale：リスケールする、大きさを変更する
re-scaling：再スケール化
reaction norm：反応基準
Reakirt's blue butterfly：メスキートヒメシジミ
real mating difference：交尾の実際の相違
real-time fluorescent quantitative PCR technique：リアルタイム蛍光定量的PCR法
realized heritability：実現遺伝率
realized niche：実現ニッチ
reappraisal：再評価
rear：飼育する
reared male：飼育された雄
rearing：飼育
rearing environment：飼育環境
rearing experiment：飼育実験
rearing temperature：飼育温度
reasonable explanation：妥当な説明、理にかなった説明
received：受け取った
receptaculum seminis：貯精のう、受精のう
receptive：感覚の、受容の
receptive female：交尾できる状態の雌
receptor gene：受容体遺伝子
receptor population：受入側個体群、受容体個体群
receptor site：受入側生息地
recessive：劣性の、劣勢形質、不顕性
recessive mutant：劣性突然変異体
reciprocal cross：正逆交配、相反交雑、相互交雑
reciprocal F1 progeny：正逆交配のF1子孫
reciprocal F1 type：正逆交配のF1型

reciprocal influence：相互作用
reciprocal pattern：相互パターン、正逆パターン
reciprocal test：相互試験、相互検定
reclivous：外斜
recognizable adult morphogenesis：認識可能な成虫形態形成
recognized sub-species：認められた亜種
recolonize：再移入する
recombinant phenotype：組み換え表現型
recombination：組み換え
recombination between color and preference：色と選好の間の組み換え
recombination breakpoint：組み換え(の)切断点
recombination distance：組み換え距離
recombination mapping：組み換えマッピング
Recommendation：勧告
reconstructed amino acid：再構成されたアミノ酸
reconstruction of nucleotide sequence：塩基配列の再構成
recorded neural activity：記録された神経活動
recording electrode：記録(用)電極
recreational purpose：レクリエーションの目的
recruitment：動員、漸増(ぜんぞう)、加入、リクルートメント、補充
rectum：直腸
recurrent pitfall：頻発する落し穴
red admiral butterfly：ヨーロッパアカタテハ
red data book：レッドデータブック（レッドリストに載せた生物を紹

介した本)
red dot：赤い丸、赤い点
red lacewing butterfly：ビブリスハレギチョウ
red light：赤色光
red list：レッドリスト（絶滅危惧種の目録）
red list category：レッドリスト区分、レッドリスト部門（レッドリストの段階分け）
red list criterium：レッドリスト基準（レッドリストの段階分けの基準）
red patch：赤い大斑点
red pigment-concentrating hormone：赤色色素濃縮ホルモン
red proboscis：赤色の口吻
red ray：赤い放射状線
red spotted pattern：赤いまだら模様
red-green color vision：赤緑色覚
reddish brown pupa：赤褐色の蛹
reduced carbon compound：希釈炭素化合物
reduced effort scientific method：軽減効果のある科学的手法
reduced fecundity：低下した生殖能力
reduced interspecific gene flow：種間の遺伝子浸透を減少させる
reduced nicotinamide adenine dinucleotide：ニコチンアミドアデニンジヌクレオチド還元型
reduced preference：弱められた選好性
reduction：縮小（図）
reduction of posterior eyespot：後方（の）眼状紋の縮小
reduplicated produced leg：過剰肢
referee：査読者、審査員、レフェリー
refereeing：審査

reference：引用文献、参照
reference pressure：基準圧、基準圧力、基準音圧
reference sequence：参照配列、基準配列、リファレンスとなるべき配列データベース
reference spectrum：基準スペクトル
refining：精製
reflectance basking：開翅日光浴
reflectance spectrum of eyeshine：暗視眼の反射スペクトル
reflected light：反射光
refractory period：無反応期、不応期
refuge area：保護区、避難区域
refuge population：保護地の個体群
refuge site：保護地、避難場所、生息地
refugia：レフュジア、待避地、待避場所
regain：回復する
regal fritillary butterfly：イダリアギンボシヒョウモン
regeneration：再生
region of reduced recombination：組み換えを減少させる領域
region sequence：領域シーケンス
regional biogeography：区系生物地理学
regional bond：地縁的
regional characteristic：地縁性
regional diversity：地域多様性
regional form：地理型
regionally rare alternative host plant：地域的にまれな代替寄主植物
regression：退化、退行、回帰式
regression line：回帰直線
regular interval：規則的な時間間隔
regularity：規則性
regulation：調節

regulatory element expansion：調節要素の拡大
regulatory gene：調節遺伝子
regulatory mechanism：調節機構
regulatory region：調節領域
regurgitation：吐き戻し
rehydrate：再水和する
reinforce：強化する、補強する
reinforcement：強化
reinsemination：再受精
reinstate：復権
reintroduce：再導入する
reintroduction：再導入
reject：拒否する、却下、リジェクト
rejected name：拒否名
rejected work：拒否された著作物
rejection behavior：拒絶行動
related compound：類縁化合物
related genera：近縁属
related species：類縁種、近縁種
relatedness：近縁性、血縁度
relatedness asymmetry：血縁度非対称性
relational：類縁関係的
relative：相対的に
relative abundance：相対的存在量、相対的な豊富さ
relative abundance distribution：相対的個体数分布
relative direction：相対方向
relative inactivity：相対的不活性
relative investment：相対的投資量
relative light intensity：相対照度
relative mating probability：相対的交尾確率
relative position：相対位置
relative probability：相対確率
relative sensitivity：相対的感受性
relative term：相対的な用語
relatively puny insect：比較的弱々しい昆虫
relay of short-range signal：短距離シグナルの中継
release：放逐、リリース、放飼
release of ecdysteroid hormone：エクジステロイドホルモンの放出
release of latent heat concomitant with freezing：冷凍で付随して発生する潜熱の放出
releaser：解発因、リリーサー、レリーサー、解発因子
releaser pheromone：解発フェロモン、リリーサーフェロモン
releasing butterfly：放蝶
relevance：関連性、妥当性
relic：名残り、遺存
relic distribution：遺存分布
relic endemism：遺存的固有
relict：遺存種、残存種
relictual butterfly：遺存種の蝶
relief pattern：レリーフパターン
relieve population pressure：個体群圧の緩和
remarkable clinal variation：注目すべきクライン（的）変異
remnant population：残存個体群
remote island：離島
removal of brain：除脳、脳除去
remove：取り除く、除去する
render：与える
rendezvous site：ランデブーサイト、面会場所
repeated hibernation：繰り返される越冬

repeated involvement：反復関与
repeated re-sampling：反復リサンプリング
repeated recruitment：反復補充
repeated recruitment of allele：対立遺伝子の反復補充
repellent volatile：忌避剤の揮発成分
replacement name：置換名
replacement reproductive：置換生殖虫
replacement substitution：変化置換、置換／代用
replanting：移植
replenish：補給する、補充する
replenishment of protein：タンパク質の補給
replicate：反復実験、レプリケート、反復、再現実験
replicated sample：重複されたサンプル
repopulate：再び居住させる
representative trace：代表的なトレース、典型的な痕跡(こんせき)
representative tree topology：典型的な系統樹形
reprint：別刷り
reproduce：繁殖する、再生する、再現する
reproduce vegetatively：栄養生殖を行う
reproduction：繁殖、生殖、増殖
reproductive aberration：生殖異常
reproductive ability：繁殖能力
reproductive activity：生殖活動
reproductive behavior：生殖行動
reproductive capacity：繁殖力
reproductive capacity of adult：成虫の繁殖能力、成虫の生殖能力
reproductive caste：生殖カースト
reproductive control：生殖制御による防除

reproductive diapause：生殖休眠
reproductive dormancy：生殖休眠
reproductive efficiency：増殖効率
reproductive effort：繁殖努力
reproductive fitness：繁殖適応度
reproductive history of adult：成虫の繁殖史
reproductive interference：繁殖干渉
reproductive isolation：生殖(的)隔離、繁殖隔離
reproductive manipulation：生殖操作
reproductive organ：生殖器官
reproductive phenomenon：生殖現象
reproductive phenotype：生殖表現型
reproductive potential：生殖能、繁殖潜在力
reproductive rate：増殖率
reproductive skew：繁殖の偏り
reproductive strategy：繁殖戦略
reproductive success：繁殖成功度
reproductive system：生殖系
reproductive tract：生殖器官
reproductive tract complex：生殖器官複合体
reproductive tract development：生殖器官の発育
reproductive tract tissue：生殖器官組織
reproductive value：繁殖価
reptile：爬(は)虫類
repulsion：撃退
repulsive activity：反発行動
reputable dealer：信頼できる業者
required heat unit：有効積算温量
rescue effect：救済効果、レスキュー効果
rescue translocation：危険から救う移動
rescuing function：救済機能

research：調査
reside：住む
residence：定住性
resident：先住者
resident dry season form individual：居住している乾季型の個体
residual：残差
residual reproductive value：残存繁殖価、残存生殖価
residual value：残差値
residue：残留物
resilin：レシリン
resistance：耐性、抵抗性
resistant：抵抗する、抵抗力がある
resonance effect：共鳴効果
resource：資源、原料
resource availability：資源の有用性、資源の利用可能性
resource budget：原料の収入分
resource defense polygyny：資源防衛型の一夫多妻
resource-holding power：資源保持力
respective co-mimics：各々の相互擬態種
respiratory system：呼吸系
respiratory volume：呼吸量
respond physiologically：生理的に反応する
response of thermoperiod：温度周期反応
response pattern：反応様式
response property：応答特性、反応特性
response rate：反応率
response system：反応系
response threshold：応答閾(いき)、応答閾値
response to current adversity：逆境に対する反応

response to selection：選択に対する反応
responsive：反応する、応答的な
rest：休む
resting point：静止位置
resting posture：休止姿勢
restoration：回復、復帰
restrain：抑制する
restraint need：抑制的な欲求
restrict：限定する
restriction fragment length polymorphism：制限酵素断片長多型、制限断片長多型（RFLP）
restriction-site associated DNA：制限酵素認識部位に関わるDNA
result：結果
resultant intense competition：結果として生じる激しい競争
resurgence：誘導多発生、リサージェンス
retain：保持する
retard：妨げる、遅らせる
retardation：遅延
retina：網膜〔ラテン語〕
retinal photoisomerase：網膜の光異性化酵素、レチナール光異性化酵素
retinal regionalization：網膜の領域化、網膜部域性、レチナール領域化
retractile tentacular organ：伸縮自在の触手状器官
retro-：逆 -、退 -〔ラテン語〕
retrograde fill：逆行性充填
retrogressive molt：退行脱皮
retrotransposon-associated coding region：レトロトランスポゾン関連コーディング領域
return：帰還する、戻る
return flight：帰還飛翔、帰還移動

return-migration：帰還移動、帰還移住、戻り移動
reveal：明らかにする
reverse gene flow：逆転遺伝子流動
reverse question：逆質問
reverse transcriptase：逆転写酵素
reverse transcription：逆転写、逆転写反応
reverse transcription polymerase chain reaction：RT-PCR法、逆転写ポリメラーゼ連鎖反応
reversible：可逆的
review：査読、総説
reviewer：査読者
revise：訂正
revised edition：改訂版
RFLP：制限酵素断片長多型、制限断片長多型（Restriction Fragment Length Polymorphism）
rhabdom：感桿（かんかん）（個眼内の光を通す円柱構造）
rhabdom waveguide：感桿の光導波路
rhabdome：感桿
rhabdomeric opsin：感桿型オプシン
rhabdomeric photoreceptor cell：感桿型光受容細胞
rhabdomeric-type：感桿型
rhinoceros beetle：オオツノカブトムシ
rhodommatin：ロードマチン
rhodopsin type：ロドプシンタイプ
rhodopsin type receptor：ロドプシンタイプ受容体
rhodopterin：ロドプテリン
Rhopalocera：蝶（チョウ）のこと（棍棒状の触角）
RI：繁殖干渉（Reproductive Interference）

ribosomal protein：リボゾームタンパク質
Rice black streaked dwarf virus：イネ黒すじ萎縮ウイルス、イネ黒すじ萎縮病
rice field：水田
rice paper butterfly：オオゴマダラ
rice plant：イネ、稲
rice skipper butterfly：イチモンジセセリ
Rice stripe virus：イネ縞葉枯ウイルス、イネ縞葉枯病
richness comparison：種数比較
richness estimation：種数推定
richness estimator：種数推定量
richness measure：種数指数
richness per quadrat：方形区当りの種数
Rickettsia：リケッチア
ridge：隆起、縦隆起、突起、稜線
ridge of morphogen：モルフォゲンの隆起線
right tympanal membrane, the：右鼓膜
ring：環
ring pattern：環パターン
ringed：リング状の
ringed spot：縁どりのある斑紋
rinse：すすぎ落とす
Rio Declaration：リオ宣言
riparian forest：渓畔林、河畔林
ripe fruit：熟した果実
ripeness：成熟
ripple pattern：さざ波状パターン
rise/fall：立ち上がり／立ち下がり
rising winter temperature：冬期の気温の上昇
risk analysis：リスク分析
risk assessment：リスク評価
risk extinction：絶滅の危機にさらす

risk management：リスク管理、危機管理
river bank：川の土手
river basin：河川流域
riverside：流域
RNAi：RNA 干渉法、RNA 干渉（RNA interference）
road kill：ロードキル、道路で車にひき殺された動物の死体
roadside：路傍
roadside butterfly：道ばたの蝶
robust：がっしりした、頑丈な
robust correlation：強固な相関
robust phylogenetic tree：堅固な系統樹
rocky area：岩場
rolled leaf：丸まった葉
room condition：室内条件
roost：とまり木
rostal：吻側
rotten fruit：腐った果実
rotting fruit：腐った果物
rough surface：粗粒面
rough-surfaced endoplasmic reticulum：粗面小胞体
round flight：往復飛翔
round Sea of Japan area：周日本海地域
rounded：丸みをおびた
RSV：イネ縞葉枯ウイルス、イネ縞葉枯病（Rice Stripe Virus）
RT-PCR：逆転写ポリメラーゼ連鎖反応（Reverse Transcription Polymerase Chain Reaction）
rudiment：痕跡的（なもの）、原基
rudiment of an eyespot：目玉模様の痕跡
rule：条項
ruling by the Commission：審議会による裁定
run transversely：横断的に延びている
rural area：田園地域
rutaceous plant：ミカン科植物
Ryukyu Islands：琉球列島

s

saccharide：糖類
sacculus：小のう、小胞〔ラテン語〕
saccus：（交尾器の）胞のう、中胞〔ラテン語〕
saddle marking：鞍状の模様
safety band：安全帯
sagittal plane：矢状断面（しじょうだんめん）、矢状面
sagittal section：矢状断面
sagittal surface：矢状断面
SAI：性非対称な近親交配（Sex-Asymmetric Inbreeding）
sal：スパルト遺伝子（*spalt*）
saline：食塩水、塩類
saline intake：ナトリウム塩の摂取、塩分摂取
same orientation：同一方向、同方向
same place：同一地方、同一場所
same species and distinct subspecies：同種異亜種
sample heterogeneity：サンプルの不均質性、サンプルの異種混交性
sample site：標本採集地、サンプル採集地、試料採集地
sample-based curve：サンプル数に基づく曲線
sample-based dataset：サンプル数に基づくデータセット
sample-based protocol：サンプル数に基づくプロトコール

sample-based rarefaction curve：サンプル数に基づく希薄化曲線
sample-size dependence：サンプルサイズ依存性
sampled site：サンプル採集地
samples：サンプル数
sampling curve：サンプリング曲線
sampling device：サンプリング方法
sampling effort：サンプリング努力、サンプリングの手数、サンプリング調査
sampling issue：サンプリング問題点
sampling protocol：サンプリングプロトコール
sampling rate：標本抽出率
sampling unit：サンプリング単位
sandwich：差し込む、間に入れる
sandy area：砂漠地帯
sap：樹液
sapling：若木
sapling dataset：若木データセット
sapling species richness：若木の種数
saprophagous butterfly：腐食性チョウ
saprophagy：腐食性
sarpedobilin：サルペドビリン、サーペドビリン
satin：サテン地
satoyama：里山
saturation of local community：局所群集の飽和
Saturn butterfly：ルリオビトガリバワモン、ルリオビトガリバワモンチョウ
Satyr butterfly：ジャノメチョウ科の蝶
satyrine butterfly：ジャノメチョウ科の蝶
Satyrs and Wood-Nymphs：ジャノメチョウ亜科
savanna：サバンナ

savanna climate：サバンナ気候
saw-tooth cline：鋸歯状のクライン
Sc：亜前縁脈（Sub coastal vein）
scaffold：足場、骨組
scale：鱗粉
scale arrangement：鱗粉の配列、鱗紛列
scale bar：目盛り棒、スケールバー
scale cell：鱗粉細胞
scale morphology：鱗粉の形態
scale mother cell：鱗粉の母細胞
scale of interest：関心度、関心度尺度
scale of space and time：時空間スケール、時空間尺度
scale row：鱗粉の列
scale row ring：鱗粉列リング
scale-forming cell：鱗粉形成細胞
scalloped edge：凸凹のある縁
scalloping：凸凹のある
scalloping flight：扇形飛行
scalloping of lateral margin：外側辺縁部の薄切り
scanning electron microscope：走査（型）電子顕微鏡
scanning electron microscopy：走査（型）電子顕微鏡検査、走査（型）電子顕微鏡法
scanning laser vibrometry：走査型レーザー振動計
scanning-electronmicrograph：走査型電子顕微鏡（像）
scape：柄節（へいせつ）
scaphium (sca)：スカヒュウム、竜骨
scar：跡
scarab beetle：スカラベ、タマコロガシ（コガネムシの仲間）
scarce：珍しい

scarce element：希少成分
scarce species：希少種
scarcity：まれなこと、稀少
scare：驚かす
scarlet：緋色(ひいろ)、深紅色
scat：糞
scattered woodland：点在している森林地帯
scavenger：腐肉食(性)動物、清掃動物
scent gland：発香腺
scent horn：臭角
scent marking behavior：匂い付け行動
scent organ：発香器官
scent patch：性紋、性標
scent sack：香のう
scent scale：香鱗、発香鱗
schematic illustration：模式図、図解図、概略図
schematic of ommatidium：個眼の模式図
scheme of society：社会機構
Schiff base：シッフ塩基
scientific evidence：科学的証拠、科学的根拠
scientific name：学名
sclerite：硬皮、節片、骨片、硬板
sclerophyll forest：硬葉樹林
scolopale cap：有桿帽鞘、有桿体帽鞘
scolopale cell：有桿体細胞、有桿細胞
scolopidia：弦状感覚子、弦音器官（scolopophore）
scolopidium：弦音器官
scolopophorous organ：弦音器官
scolus：(複.-lus)、棘瘤、枝刺〔ラテン語〕
scotophase：暗期
scotophil：親暗相
SCP：過冷却点(SuperCooling Point)

scraping：こする音、削る音
screen house：網室
screen out：スクリーニング、ふるい分け
screened lid：スクリーン付き蓋
scrub：潅木、雑木林、こする
scrub clearance：雑木林の撤去
scrubland：低木林地帯
SE：標準誤差(Standard Error)
sea level：標高、海水面
Sea of Japan Rim Area：周日本海地域
search image：探索像
searching image：探索像
season：時期、季節
seasonal adaptation：季節適応
seasonal arrangement：季節配置
seasonal change：季節的変化、季節(的)変動
seasonal change of host plant：寄主植物の季節的転換
seasonal cycle：季節周期
seasonal decline：季節的低下、季節的減少
seasonal difference in fitness：適応性の季節的相違
seasonal dimorphism：季節的変異
seasonal diphenism：季節性二型
seasonal environment：季節(的)環境
seasonal form：季節型
seasonal form response：季節型反応
seasonal habitat variability：生息地の季節的変動性
seasonal index：季節指標
seasonal information：季節情報
seasonal migration：季節的移住、季節移動性
seasonal morph：季節型
seasonal morph response：季節型反応

seasonal periodicity : 季節周期
seasonal polymorphic migrant : 季節多型性移動
seasonal polymorphism : 季節(的)多型
seasonal polyphenism : 季節多形、季節的表現多型、季節の多型
seasonal rhythm : 季節的リズム
seasonal variability : 季節的の変異性
seasonal variation : 季節(的)変異
seasonal wing-morph : 季節的翅型
seasonal-form determination : 季節型決定
seasonal-form expression : 季節型発現
seasonal-form-determining hormone : 季節型決定ホルモン
seasonal-morph determination : 季節型決定
seasonal-morph expression : 季節型発現様式
seasonal-morph variation : 季節型変異
seasonality : 季節性
second generation : 第二世代
second hypothesis : 第二仮説
second instar : 二齢
second stainless steel electrode : ステンレススチール製二次電極
second-growth forest : セカンドグロース林、二次成長林
secondary attractant : 二次誘引物質
secondary compound : 二次代謝産物
secondary contact : 二次的接触
secondary forest : 二次林、二次森林
secondary homonym : 二次同名、二次異物同名
secondary metabolite : 二次代謝産物
secondary plant substance : 二次植物成分
secondary seta : 二次刺毛

secrete : 分泌する
secretion : 分泌(物)、分泌状態
secretory granule : 分泌果粒
secretory protein : 分泌タンパク質
section : 節、薄片を作る
section longitudinally : 縦方向に切断する
section transversely : 横方向に切断する
sectoral crossvein : 径分横脈、s脈
sediment : 堆積物
sedimentary rock : 堆積岩
seed : 種子
seed pod : 種子のさや、種鞘、シードポッド
seed set : 結実率、種子形成
segment : 節、体節、セグメント、分節
segregate : 分離する、隔離する
segregation : 分離
selected site : 選択された部位
selection : 選択、淘汰、選抜
selection coefficient : 選択係数
selection differential : 選択差
selection favoring single-locus control : 一遺伝子座の支配を有利にする選択
selection pressure : 選択圧、淘汰圧
selective advantage : 淘汰利益、淘汰の有利性、選択有利性、選択優位性、選択的優勢
selective disadvantage : 淘汰不利益、淘汰の不利性、選択不利性
selective factor : 選択要因
selective felling : 選択(的)伐採
selective force : 選択圧、淘汰圧
selective harassment : 選択的ハラスメント、選択的干渉

selective logging：選択(的)伐採、択伐林業
selective pressure：選択圧、淘汰圧
selector gene：選択遺伝子
self-assembly：自己組織化
self-incompatibility：自家不和合性
self-matching：自分と釣り合う(相手との交尾)
self-organization：自己組織化
self-pollination：自家受粉
selfing-avoidance：自己回避
selfish element：利己的因子
selfish endosymbiont：利己的な細胞内共生微生物
selfish genetic element：利己的遺伝子
semel-：一度-、一回-〔ラテン語〕
semelparity：一回繁殖
semelparous：一回繁殖
semelparous colonizer：一回結実性の移住種、一回繁殖性の移住種
semi-：半-〔ラテン語〕
semi-dominant mutation：半優性突然変異
semi-natural grassland：半自然草地
seminal fluid substance：精液物質
seminal protein：精液タンパク質
seminal vesicle：貯精のう
seminatural habitat：半自然の生息地
semiochemical：信号物質、信号化学物質
semisocial route：半社会性ルート、同世代個体共存
semispecies：半種(形態的な差異は認められるが、生殖隔離がなされていない)
semivoltine：半化性(生活史が二年がかり)
senior author：シニアオーサー、論文指導者、首席著者、主席著者

senior homonym：古参同名、上位異物同名
senior synonym：古参異名、上位同物異名
sense organ：感覚器
sensillum：(複.-la)感覚子、感覚器〔英語〕
sensillum styloconica：歯状感覚子
sensitive period：敏感期
sensitivity：感受性
sensory：感覚
sensory cell：感覚細胞
sensory cell body：感覚細胞体
sensory hair：感覚毛
sensory innervation：感覚神経支配
sensory mechanism：感覚機構
sensory process：感覚過程
sensory structure：感覚器官の構造、感覚構造
sensu：の意味で〔ラテン語〕
sensu lato：広義の、広い意味で〔ラテン語〕
sensu stricto：狭義の〔ラテン語〕
separate：抜き刷り
separate species：独立種
separation zone：分離帯
sepiapterin：セピアプテリン
septum：隔膜
sequence alignment：配列アラインメント、配列アライメント
sequence divergence：配列多様性、塩基配列間の相違、配列間の差異
sequence of mitochondrial DNA：ミトコンドリア DNA の塩基配列
sequence the genome：ゲノムの塩基配列を解読する
sequenced species：遺伝子配列された種
sequenced taxa：遺伝子配列された分類群
sequestration：蓄積、滞留
sequestration of flavonoid：フラボノイド

の蓄積
series of ocelli：眼状紋列
serine：セリン
serine-to-alanine substitution：セリンからアラニンへの置換
serious decline：深刻な衰亡
serosa：漿膜（しょうまく）
serpentine soil environment：蛇紋岩地の土壌環境
seta：（複. -ae)，刺毛，剛毛〔ラテン語〕
settle：定住する
settling contest：決着済みの競争
seven transmembrane protein：7回膜貫通型タンパク質
sever：切断する
severe competition：激しい競争
severe long-term decline：厳しい長期間にわたる衰亡
sex brand：性標（斑）
sex chromatin body：性染色質体
sex chromosome：性染色体
sex determination：性決定
sex determination and differentiation：性決定と性分化
sex difference：性差
sex lethal gene：性致死遺伝子
sex mosaic：雌雄モザイク、性モザイク
sex patch：性標
sex pheromone：性フェロモン
sex pheromone formation：性フェロモン形成
sex pheromone unit：性フェロモン単位
sex ratio：性比
sex-asymmetric inbreeding：性非対称な近親交配
sex-differentiating hormone：性分化ホルモン
sex-limited inheritance：限性遺伝
sex-linkage：伴性、性連鎖、伴性遺伝
sex-linked：伴性（の）、伴性遺伝の
sex-linked diapause response：伴性の休眠反応
sex-linked gene：伴性遺伝子
sex-linked inheritance：伴性遺伝
sex-related gene：性関連遺伝子
sex-specific mRNA splicing：性特異的なmRNA スプライシング
sexual communication：性的通信
sexual difference：性差、性的差異
sexual dimorphism：性的二型（性）
sexual harassment：性的干渉、セクシャルハラスメント、性的いやがらせ
sexual isolation：性的隔離、生殖（的）隔離
sexual marking：性標
sexual mosaicism：性的モザイク現象
sexual reproduction：有性生殖
sexual selection：性選択、性淘汰
sexually dimorphic wing morphology：性的二型の翅形態
sexually intermediate genitalia：性的に中間の交尾器
sexually intermediate phenotype：中間的な性表現型
sexually intermediate trait：性的に中間型の形質、間性形質
sexually mosaic individual：性モザイク個体
Sf9：ツマジロクサヨトウ（*Spodoptera frugiperda*）の卵巣細胞由来の樹立培養細胞株
shadow hindwing bar：後翅線条の透視画像

shake：振盪(しんとう)する
shape：形、形状
shaped pulse：整形パルス、形状パルス
shared developmental pathway：共通の発育経路
shared genetic architecture：共通の遺伝的構造
shared sequence variation：共有配列変異、共通配列変異
sharply pointed：鋭くとがった
shear：切断する
sheath：鞘(さや)
shed：与える
shed light on：光を当てる
sheep：ヒツジ、羊
shell：殻
shelter：隠れ家、シェルター
shelter-dwelling caterpillar：隠れ家で暮らしている幼虫
shift of voltinism：化性転換
shifting balance：平衡遷移、平衡推移、バランス変動、推移平衡
shivering：シバリング、粉砕、はく裂
short day length：短い日長
short hair：短毛
short latency：短潜時
short period of time：短時間、短期間
short photoperiod：短日日長
short turf：丈の短い芝
short-day effect：短日効果
short-day frequency：短日サイクル数
short-day photoperiod：短日日長
short-day treatment：短日処理
short-day type：短日型
short-day type response：短日型反応
short-focus telescope：短焦点の望遠鏡

short-horn type：短角型(たんかくがた)
short-wavelength-sensitive cone pigment：短波長感受性錐体視物質
shotgun sequencing method：ショットガンシークエンス法
shrink：小さくなる、縮む
shrub：低木、灌木(かんぼく)
shrubbery：低木、生け垣
SI：群知能(Swarm Intelligence)
sib mating：同胞交配、近縁交配、兄弟交配
Siberian distributed type：シベリア分布系統
Siberian type：シベリア型
sibling group：近縁グループ、兄弟グループ
sibling species：同胞種、近縁種
sieving：ふるい分け、ふるい操作
sight：視覚
signal molecule：シグナル分子
signal peptide：シグナルペプチド
signal preference：シグナル選好
signal production：シグナル生産
signal reception：シグナル受容
signal transduction：シグナル伝達、シグナル伝達系
signaling factor：シグナル因子
signalling：シグナル伝達、信号伝達
signalling pathway：シグナル伝達経路、シグナル伝達系
signalling-pathway gene：シグナル伝達系遺伝子
signature of selection：自然淘汰のサイン、自然淘汰のしるし
significance threshold：有意な閾(いき)値、有意性閾値

significant deviation：有意な偏差
significant difference：有意な相違
significantly less：有意により少ない
significantly positive D-statistic：有意な正のD統計量
significantly prefer：有意に好む
significantly worse：有意に悪い、大幅に劣る
silicon grease：シリコーングリース、シリコングリス
silk：絹糸
silk girdle：帯糸（たいし）
silk gland：絹糸腺
silken girdle：絹の支持帯
silkworm：カイコガの幼虫、ヤママユガ科の幼虫の総称
silvaniform clade：シルヴァーニ型の分岐群
silver hairstreak butterfly：シラメスアカミドリシジミ
silver-spotted skipper butterfly：ホクベイオオギンモンセセリ、ホソバセセリ
silvicultural system：造林システム、造林施業システム
similar compound：類似化合物
similar species：近似種
similarity：類似、類似性、類似度
simple eye：単眼
simple family：単純家族
sine wave：サイン波、正弦波
single autosomal locus：単一の常染色体遺伝子座
single butterfly genus：蝶の単一属
single clone：単一クローン
single diapause stage：一回限りの休眠期、単発休眠期

single gene mutant：単一遺伝子の突然変異体
single gene mutation：単一遺伝子の突然変異
single genotype：単一（の）遺伝子型
single lens reflex camera：一眼レフカメラ
single locality：単一局所性
single nucleotide site：一塩基部位、一ヌクレオチド部位
single nucleotide transition difference：一塩基の転位差
single optical-sensing structure：単光センシング構造
single origin：単一起源
single positive clone：単一陽性クローン、単一ポジティブクローン
single round only：一回限り
single S opsin variant：単一S（短波長型）オプシン変異体
single sensillum recording：単一感覚子記録法
single spectral class of photoreceptor：光受容器の単一スペクトル型
single switch gene：単独スイッチ遺伝子
single-brooded：一化の
single-copy microsatellite：単コピーマイクロサテライト
single-copy nuclear gene：単コピー核遺伝子
single-copy nuclear locus：単コピー核遺伝子座
single-locus inheritance：単一遺伝子座の遺伝
single-strain-infected normal brood：単系統に感染した正常な同腹仔
single-stranded sequencing reaction：一本

鎖シークエンス反応
singleton：1個体だけ現れた種、単一個体（種）
singly infected female：単感染の雌
singular point：特異点
sinigrin：シニグリン
sink habitat：シンク生息地（繁殖はできるが個体数が減少してしまう場所）
Sino-Japanese region：日中区、日中区系
Sino-Japanese type：日中系統、日中型
Siphonaptera：ノミ目、隠翅目
sire：子供を作る
sister genus：姉妹属
sister group：姉妹群
sister haplotype：姉妹ハプロタイプ
sister species：姉妹種
sister taxon：姉妹分類群、姉妹群
SIT：不妊虫放飼法（Sterile Insect Technique）
site-specific scale：現地に特有の規模
sitting behavior：静止行動
sitting time：静止時間
situate：置く、位置を定める
size：大きさ、寸法
size of a potential inversion：潜在的な逆位のサイズ
size of sample：サンプルの大きさ、サンプルサイズ
size variation：サイズ変異
skeletal soil：粗骨土壌
skeleton：骨格、枠、スケルトン
skeleton photoperiod：スケルトン光周期、枠光周期
skew：偏り
skill of searching literature：文献探索技能
skin：表皮、クチクラ

skipper butterfly：セセリチョウ科の蝶
skippers, the (superfamily Hesperioidea)：セセリチョウ上科
skull：頭蓋骨
slender：細い
slender hair：細い毛
slice preparation：薄切標本
slide glass：スライドガラス
sliding-window phylogenetic analysis：スライディングウィンド手法を用いた系統解析
slight local variation：微妙な局所的変異
slightly distal：わずかに末端へ、わずかに末梢へ
slightly elliptical eyespot：少し楕円形状の眼状紋
Sloane's urania moth：スロアヌスオオツバメガ（スローン氏に因んで命名）
slope：坂、斜面、傾き
slug-like：ナメクジのような、ナメクジ形の
small postman butterfly：アカスジドクチョウ
small tortoiseshell butterfly：コヒオドシ
small-scale map：小縮尺の地図
small-scale rearrangement：小規模再配置
smell：嗅覚、におい
smooth cline：滑らかなクライン
smooth curve：滑らかな曲線
smooth surface：平滑な面
smoothed rarefaction curve：滑らかな希薄化曲線
snap-shot camera：コンパクトカメラ
sneaker：スニーカー
snout：鼻
snout butterfly：テングチョウ

Snouts：テングチョウ亜科
social communication：社会的コミュニケーション
social ecology：社会生態学
social parasitism：社会寄生
sociality：社会性
socio-economic：社会経済的な
socket：ソケット
socket-forming cell：ソケット形成細胞
sodium：ナトリウム
sodium carbonate：炭酸ソーダ、炭酸ナトリウム
sodium-supplemented mud puddle：ナトリウムが補給された泥水
soft leaf：柔らかい葉
soil：地面、土壌
soil formation：土壌形成
soil ingestion：土壌摂取
soil surface：地表面、土壌表面
solar collector：太陽熱集熱器
solar compass：太陽コンパス
solar heat：太陽熱
solid bar：黒色の棒
solid dot：黒丸、黒点
solid-line：実線
solitary：単独性
solubility：溶解度
solvent extraction：溶媒抽出
somatic component：体成分
somatic maintenance：身体の維持
sorbitol：ソルビトール
soret band：ソーレー帯
sound level meter：騒音計、音圧レベル計
sound production：音発生
sound production behavior：発音行動
sound pulse：パルス音、音(波)パルス

sound stimulus：音刺激
sound-producing：音を発生する
source：ソース、(供給)源、供給側
source country：原産国
source habitat：ソース生息地、ソースハビタット(繁殖に好適で子孫が増える場所)
source population：供給(元)個体群、供給側の個体群
source-sink model：ソース - シンクモデル、本土 - 島モデル
southerly flight：南方へ飛翔
southern dogface butterfly：ミナミイヌモンキチョウ
southern marginal area：南縁部
Southern rice black streaked dwarf virus：イネ南方黒すじ萎縮ウイルス、イネ南方黒すじ萎縮病
southern species：南方性の種、南方的な種
southwest direction：南西方向
southwest lowland of Japan：南西日本の低地、日本の南西地方の低地
southwestern island population：南西諸島の個体群
soyasaponin：大豆サポニン
sp：スパルト遺伝子(*spalt* gene)
sp.：種(species)
sp. aff.：「種が確定していない時にとりあえずある種に類似していることを表す(species with affinity)」、関係種、近似種
sp. nov.：新種名(species novus)
space：翅室
space out：間置き
spalt：スパルト遺伝子
span：指寸法で計る、架かる

sparse vegetation：点在する植生
spatial aggregation of species：種の空間的集団
spatial anatagonism：空間拮抗性
spatial and temporal population dynamics：時空間の個体群動態
spatial arrangement：空間配列
spatial autocorrelation：空間的自己相関
spatial configuration of suitable habitat：好適な生息地の空間配置構造
spatial distribution of habitat：生息地の空間分布
spatial distribution of individual：個体の空間分布
spatial information：空間情報
spatial pattern：空間分布パターン
spatial positioning：空間位置決め
spatial scale：空間スケール、空間尺度
spatial structure：空間構造
spatial unit：空間単位
spatial variation：空間的変異、空間変異
spatially uncorrelated stochasticity：空間的に相関関係のない確率性
speaker：スピーカー
special protection zone：特別保護地区
specialist：スペシャリスト、専食者、専門家
specialist species：スペシャリスト種
specialization：特化、分化
specialized：分化した
speciate：種分化する
speciation：種分化、分化
speciation date：種分化の時期
speciation history：分化の歴史
species：種（分類階級の「しゅ」）
species accumulation curve：種数累積曲線

species boundary：種の境界
species conversion：種の転化
species density：種密度
species distribution：種の分布
species diversity：種多様性
species endemic to Japan：日本特産種
species extinction：種の絶滅
species group：種階級群
species hibernating as egg：卵休眠種
species hibernating as pupa：蛹休眠種
species inquirenda：（複．-ae）、未確定種
species level：種レベル
species loss：種の喪失
species name：種名
species name list of Japanese butterflies：日本産蝶種名一覧
species per genus：属あたりの種数
species recognition：種認識
species richness：種数、種の豊富さ
species selection：種選択
species specific：種特異的
species' extinction：種の絶滅
species--family ratio：種数 - 科数の比
species-abundance model：種数 - 個体数モデル
species-group name：種階級群名
species-per-individual ratio：個体数あたりの種数の比
species-poor：種が少ない、種が乏しい
species-poor community：種が少ない群集
species-rich：種が多い、種が豊富な
species-rich clade：種類が多い単系統
species-rich community：種が多い群集
species-rich genus：種数の多い属
species-specificity：種特異性

species-to-genus ratio：種数対属数の比
species-to-individual ratio：種数対個体数の比
species/individual ratio：種数／個体数の比
specific cue：固有の刺激
specific mixture of secretion：分泌物の特定混合物、分泌物の特異的混合物
specific name：種小名
specific nature of the association：結合の特異的な性質
specific site：特定部位、特殊部位
specific sound frequency：特定音周波数
specification of focus：フォーカスの仕様、焦点の内訳
specify：指定する、規定する、特定する
specimen：標本
speciose：種数が多いこと
speck change：斑紋変化
spectacular array：壮観な配列
spectacular difference：目を見張る違い
spectral difference：スペクトル差
spectral diversification：スペクトル多様化
spectral phenotype：スペクトル表現型
spectral receptor type for color vision：色覚用のスペクトル受容体の型
spectral reflectance：スペクトル反射率、分光反射率
spectral sensitivity：分光感度、スペクトル感受性、スペクトル感度、分光応答度
spectral tuning：スペクトル同調、分光学的微調整
spectral tuning effect：スペクトル調整効果
spectral tuning site：スペクトル調節部位
spectral variation：スペクトル変化、スペクトル変動

spectrally distinct class of photoreceptor：光受容器のスペクトル的に異なる型
spectrally shifted lineage：スペクトル的にシフトされた系統、スペクトル偏移系統
spectrophotometrical characteristic：分光学的特徴
spectrophotometrically：分光学的に
speculate：熟考する、推測する、見当をつける
speculation：推測、憶測
speculative：理論的な、推論にすぎない、思いつき程度の
speculum：検鏡〔ラテン語〕
speed：飛翔速度
speed of fluttering：飛翔速度
SPEEDI：緊急時迅速放射能影響予測ネットワークシステム（System for Prediction of Environmental Emergency Dose Information）
spelling：綴(つづ)り
sperm：精子(せいし)
sperm competition：精子(間)競争
sperm count：精子数
sperm precedence：精子優先度、精子の優先性
sperm production：精子の生産
sperma-：精子-〔ギリシャ語〕
spermatheca：受精のう
spermatogenesis：精子形成
spermatophore：精包(せいほう)、精球
spermatophore protein：精包タンパク質
spheroidal wing surface：回転楕円体状の翅面
spherulocyte：小球細胞（昆虫血球細胞の一種）

sphingophilous flower：虫媒花
sphinx moth：スズメガ科の蛾
sphragis：スフラギス、受胎のう、交尾のう
spicular：針状
spider：クモ
spin：糸を吐く
spine：トゲ、刺、棘(とげ)
spinneret：吐糸管
spiracle：気門(きもん)
spiral flight：らせん飛翔、卍ともえ飛翔
spiraling：らせん飛翔
Spiroplasma：スピロプラズマ
SPL：音圧レベル(Sound Pressure Level)
split：分岐する
Spodoptera frugiperda：ツマジロクサヨトウ
spontaneous feeding response：自発的な摂食反応、自然発生的な摂食反応
spontaneous mutant：自然誘発突然変異体、自然に起こる突然変異体
spontaneous single-gene mutant：自然(に起こる)単一遺伝子突然変異体
sporadic：散発的な
spot：斑点
Spotty：斑点のある
spp.：「『その属の生物を複数種確認したが，種名は同定できなかった』ことを表している("sp."の複数形； "species"の略語)」、「その属の複数の種を指す表記方法」
spp. nov.：複数個の新種名(species novus)
spray：(殺虫剤などの)噴射
spread northward：北方へと広がる
Spread-wing Skippers：チャマダラセセリ亜科
spreading：展翅(てんし)
spreading board：展翅板
spreading fresh specimen：生展翅
spreading relaxed specimen：軟化展翅
spring ephemeral：早春季生物
spring form：春型
spring morph：春型
spring remigration：春季の再移動
spring-form female：春型雌
spring-form induction：春型誘導
SPU：性フェロモン単位(Sex Pheromone Unit)
spur：距(きょ)、蹴爪(けづめ)、端刺(たんし)
Spurr's epoxy resin：スパーのエポキシ樹脂
sputter-coated：スパッタリングコートした、スパッタにて被覆した
squarish：角張った
squeezed out of, be：追い出される、閉め出される
squid photoreceptor cell：イカ光受容体細胞
squid retinochrome：イカ網膜クロム、イカロドプシン
SRBSDV：イネ南方黒すじ萎縮ウイルス、イネ南方黒すじ萎縮病(Southern Rice Black Streaked Dwarf Virus)
Ss：歯状感覚子(Sensillum styloconica)
ssp.：亜種(subspecies)の略語、「『属名 種名 ssp. 亜種名』の形式で表記」
ssp. nov.：新亜種名(subspecies novus)
sspp.：亜種("ssp."の複数形； "subspecies")の略語
stability effect：安定効果
stability mechanism：安定機構

stabilization of national distribution：国レベルの分布の安定化
stabilizing selection：安定化選択、安定性選択、安定化淘汰
stable asymptote：安定的漸近
stable environment：安定した環境
stable isotope experiment：安定アイソトープ実験、安定同位体実験
stable landscape heterogeneity：安定した景観の異質性
stable polymorphism：安定多型
stadium：齢、期
stage：ステージ、期、段階
stain：着色する、染色する
stainless steel hook electrode：ステンレス鋼製フック電極、ステンレススチール製フック電極
stand：林分
standard curve：標準曲線
standard deviation：標準偏差
standard rearing temperature：標準的な飼育温度
standard sample：スタンダードサンプル、標準的サンプル
standardized comparison：標準的比較、標準化された比較
standpoint of phylogeny：発生的な立場、系統的な立場
starch：デンプン
startle：威嚇する
startling eyespot：脅かしの目玉模様
starvation：絶食、飢餓
starvation endurance：餓死耐久性
starve：餓死する
starve to death：餓死する
state of origin：原産国

stationary molt：定常脱皮
statistical analysis：統計解析
statistical comparison：統計的比較
statistical expectation：統計的期待値
statistical significance：統計的有意性
statistical study：統計学的研究
statistical universe：統計母集団、統計集団
statistically significant：統計的に有意な
steam distillation：水蒸気蒸留法
steep decline：急激な衰亡
stem (of a name)：語幹（学名の）
stem：枝、茎、幹、止める
stem density：樹幹密度、幹数密度、本数密度
stem mother：幹母
stem of leaf：葉柄
stemma：（複．-ata)、点眼（完変態の幼虫の頭側部にある個々の眼）、単眼
steppe environment：ステップ的環境
stepping stone habitat：足がかりの生息地、踏み石の生息地
stepwise analysis of deviance：逐次的逸脱度分析、逸脱度逐次分析
stepwise linear regression：逐次線形回帰
stereo microscope：実態顕微鏡、実体顕微鏡
stereo power amp：ステレオパワーアンプ
sterile caste：不妊カースト
sterile insect technique：不妊虫放飼法
sterility：不妊性、不稔性
sternite：腹板
sternum：腹板
stigma：（複．-ata)、性標（斑）、縁紋、柱頭〔ラテン語〕
stigmergy：スティグマジー

143

still-steep sample-based rarefaction curve：急傾斜型のサンプル数に基づく希薄化曲線
stimulant：刺激因子
stimulate：刺激する、誘導する
stimulus duration：刺激持続期間
stimulus effect：刺激効果
stimulus intensity：刺激強度
sting：刺針
stinging hair：刺毛
stink club：悪臭を放つ棍棒
stipple：点々を付ける
stochasticity：確率性
storage protein：貯蔵タンパク質
straight white band：一文字状白帯
strain：系統、株、菌株
strataum：層
strategy：戦略
stratified random design：層化無作為抽出法
stray：道に迷う
stray butterfly：迷蝶（めいちょう）
stray light：迷光
stray species：迷蝶（めいちょう）、迷い種
streak：条
stream：小川、流れ
stream assemblage：水生生物群集
stream invertebrate：小川に生息する無脊椎動物、河川性無脊椎動物
Strepsiptera：ネジレバネ目、撚翅目
stretch out：広げる
stretch receptor：伸張受容器
stridulation：摩擦発音
striking：派手な
striking variation：著しい変動
stripe：縞

striped policeman butterfly：シロオビアフリカアオバセセリ
striping：縦縞模様
strong daylight：昼間の強い自然光
strong developmental constraint：発生上の強い制約、成育上の強い制約
strong dominance：強力な優性
strong magnetic field：強磁場
stronghold：生息地
strongly suggest：強く示唆する
structural：構造色である
structural color：構造色
structure determination：構造の決定
structure-activity relationship：構造活性相関
study species：研究対象の種
stylet：口針
Styrofoam plank：スタイロフォーム厚板（発砲スチロールの厚板）
sub coastal vein：亜前縁脈、Sc 脈
sub-：亜 -、下 -〔ラテン語〕
subadult：終齢、終令
subalpine zone：亜高山帯
subapical：（翅）亜頂室、（翅）亜端室、亜頂端の、亜翅頂部、亜翅頂部帯、亜翅端部
subapical cell：亜頂室、亜端室
subarctic species：亜北極圏種
subarctic zone：亜寒帯
subclass：亜綱（あこう）
subcosta：亜前縁脈（あぜんえんみゃく）、Sc脈
subcostal vein：亜前縁脈
subdiscal line：内横線
subdorsal：亜背域
subfamily：亜科
subfamily name：亜科名

subg. : 亜属（subgenus）の略語
subgeneric name : 亜属名
subgeneric speciation : 亜属的分化
subgenual organ : 膝下（しっか）器官
subgenus : 亜属
subgenus name : 亜属名
subjective : 主観的
subjective synonym : 主観同物異名、主観異名
sublateral : 下外側
submarginal : 亜外縁部
submarginal band : 亜外縁の縞状バンド
submedial : 亜中央帯
submitted : 投稿中
suboesophageal ganglion : 食道下神経節、食道下神経球
suborder : 亜目
subordinate taxon : 従属的タクソン
subpopulation : 部分個体群（【生態学】）、サブ個体群、分集団、部分母集団（【統計学】）
subsequent analysis : その後の解析
subsequent designation : 後指定、その後の指示
subsequent monotypy : 後世の単型
subsequent spelling : 後綴（つづ）り
subsidence : 沈降、沈下、陥没
subsocial : 亜社会性
subsocial route : 亜社会性ルート、母娘共存
subsp. : 亜種（subspecies）の略語
subspecies : 亜種
subspecies name : 亜種名
subspecific name : 亜種小名
subspp. : 亜種（"subsp." の複数形；"subspecies" の略語）

substantial change : 実質的な変化
substitute name : 代用名
substitution : 置換
substrate : 基質
subtaxa : 細分類
subtaxon--taxon ratio : 細分類数 - 分類数の比
subtle climatic : 敏感な気候
subtribe : 亜族
subtribe name : 亜族名
subtropical : 亜熱帯
subtropical area : 亜熱帯地域
subtropical district : 亜熱帯地域
subtropical group : 亜熱帯グループ
subtropical population : 亜熱帯の個体群、亜熱帯性個体群
subtropical species : 亜熱帯性の種
suburbs of Tokyo : 東京近郊
subzero-temperature : 零下の気温、氷点下の気温
successful copulation : 交尾成立
successfully emerged adult insect : 首尾よく羽化した成虫
succession : 遷移、継承、サクセッション
succession of host plant : 寄主植物の遷移
successive instar : 連続的な齢期
succulent plant : 多肉植物
suck nectar : 吸蜜する
sucker : ひこばえ、吸盤、吸器
sucrose : ショ糖、スクロース
sucrose intake : ショ糖摂取
sucrose solution : ショ糖溶液、スクロース溶液
sudden dash : 急激な突進
suffix : 接尾語
sugar solution : 砂糖溶液

sugar-alcohol：糖アルコール
suitable breeding habitat：好適な繁殖地
sulcus：(複.-ci)、溝〔ラテン語〕
Sulphurs：モンキチョウ亜科
summary：摘要、要約
summative：加算的
summer diapause：夏眠、夏休眠
summer form：夏型
summer form adult：夏型成虫
summer form induction：夏型誘導
summer morph：夏型
summer-form female：夏型雌
sun compass：太陽コンパス
sun's azimuth：太陽方位
sunflower：ヒマワリ
sunlight intensity：太陽光の強度、太陽光の強さ
sunset：日没
sunspot：日溜り
super gene：超遺伝子
super-：上 -〔ラテン語〕
supercool：過冷却
supercooling point：過冷却点
superfamily：上科(じょうか)、超科
superfamily name：上科名
superficial：浅(せん)
supergene：超遺伝子、スーパー遺伝子
supergene architecture：超遺伝子構成
superior：上部の、上(方)の、上(じょう)〔ラテン語〕
supernumary eyespot：過剰に形成された眼状紋
supernumerary limb：過剰肢
superorder：上目
supplementary feeding：補助的摂食
supplementary queen：補充女王

supplementation：補強、強化
supply：供給量
suppressed infection density：抑制された感染密度
suppressed name：抑制名、抑圧名
suppressed work：抑制された著作物
suppression：抑制、抑圧
suppression factor：抑制要因
suppression of ovarian development：卵巣の発達停止
supra-：上の -、超えた -〔ラテン語〕
suprageneric：属よりも高位の
supralateral：上外側
surface area：表面領域
surface structure：表面構造
surface topography：表面形状
surgical experiment：外科的実験
surgical manipulation：外科的操作
surplus stock：過剰ストック、余剰ストック
surround：辺縁、周辺、周囲
survival：生き残り、生存
survival rate：生存率
survival value：生存価、生存上の利益
surviving progeny：生存する子孫
survivor：生き残り、遺存種
survivorship curve：生存曲線
susceptibility：感受性
sustainable development：持続可能な開発
sustainable woodland management：持続可能な森林地帯の管理
sustenance：食物源
suture：縫合線(ほうごうせん)
swallowtail butterfly：燕尾系アゲハチョウ、尾状突起付きアゲハチョウ
Swallowtails：アゲハチョウ亜科、アゲ

ハチョウ科
swamp：湿地、沼地
sward：草地
swarm：群飛、昆虫の群れ
swarm intelligence：群知能
swarm robot：群ロボット
sweetish substance：甘露
swidden：焼畑
switch supergene：スイッチ超遺伝子
Switching Mechanism At 5'-end of the RNA Transcript：SMART 法
sword-grass brown butterfly：ベニモンクロヒカゲ
sword-like tail：剣状の尾状突起
symbiont density regulation：共生微生物密度の調節
symbiont infection：共生微生物感染
symbiont-depleted insect：共生微生物が激減した昆虫
symbiont-induced reversal of insect sex：共生微生物による昆虫の性転換
symbiosis：(複 .-ses)、共生〔ラテン語〕
symbiote：共生者
symbiotic association：共生関係
symbiotic bacteria：共生細菌
symbiotic microorganism：共生微生物
symmetrical wing damage：対称的な翅の損傷
symmetry system of band：縞状バンドの相称系、相称系の縞状バンド
symmetry system of colored band：着色された縞状バンドの相称系
sympatric：同所性の、同所性
sympatric allele：同所的対立遺伝子
sympatric co-mimetic postman butterfly race：同所的相互擬態型ベニモンドクチョウの亜種
sympatric distribution：同所的分布
sympatric form：同所的変異型
sympatric mimetic taxon：同所的擬態分類単位
sympatric morph：同所的形態
sympatric race：同所的系統
sympatric speciation：同所的種分化
sympatric species：同所(的)種
sympatrically：同所的に
sympatry：同所性
symplesiomorphy：共有祖先形質、祖先形質共有
symptom：病徴、症状
syn-：共に -、同時に -、類似 -、合成の -〔ギリシャ語〕
syn.：同物異名(synonym の略語)、シノニム、異名
synapomorphy：共有派生形質、派生形質共有
synchronic：共時性、共時的な
synchronicity：共時性(形態種：形態的特徴で区別された種)
synchronized biennial butterfly：同期化した二年生の蝶
synchronous：斉一の、同時(性)の
synchrony：同時発生
syndrome：症候群、形質群、シンドローム
synephrine：シネフリン
synephrine receptor：シネフリン受容体
synergist：協力者
synergistically：相乗的に
synomone：シノモン
synonym：同物異名、シノニム、異名
synonymic note：同物異名の検討
synonymous change：同義置換

synonymous substitution：同義置換
synonymy：異名表、異名関係、異名リスト
synteny：シンテニー（同一の染色体に数個の遺伝子がのっていること）
synthesis of ecdysteroid hormone：エクジステロイドホルモンの合成
synthesize：合成する
syntype：シンタイプ、等価基準標本
system of nature：自然の体系
systematics：分類学、体系学、系統学

t

t-test：t 検定
tactics：戦術
tactile sense：触覚
tactile seta：触知性刺毛、触刺毛
tactile signal：触覚刺激
tagging：タギング、タグ付け
taiga：タイガ、北方針葉樹林
tail：尾状突起、尾
tail photoreceptor：尾端光受容器
tailless：尾状突起を欠く
tailwind drift：追い風ドリフト、追い風偏流
tall grassland：背の高い草地
tally：集計する、一致する、符合する
tandem duplication：縦列重複
tannin content：タンニン含量
tanning：硬化・着色（黒く硬化する）
tansy extraction：ヨモギギク抽出
tap water：水道水
taper off：次第に減る
tapered：だんだん細くなる
tapetum：タペータム、反射層板、内面層、タペート
target organ：標的器官
target PCR product：目的の PCR 産物
target species：対象種
tarsal claw：脚ふ節の鉤爪
tarsal contact chemosensory hair：ふ節の接触化学感覚毛
tarsus：（複 .-si）、跗節（ふせつ）〔ラテン語〕
taste：味覚、味見
taste sensillum：味覚感覚子
taste-enhancing effect：味覚増強効果
tautonymous name：反復名
tautonymy：同語反復
taxa group：分類群
taxon：（複 .-xa）、タクソン、分類単位、分類学的単位、分類群〔ラテン語〕
taxon density：分類密度
taxon occurrence：分類出現頻度
taxon richness：分類数
taxon sampling：分類サンプリング
taxon sampling curve：分類サンプリングの曲線
taxon-rich group：分類群が多いグループ
taxonomic boundary：分類学的境界
taxonomic designation：分類学的名称、分類学上の名称
taxonomic distribution：分類学的分布
taxonomic field：分類分野
taxonomic group：分類（学的）群
taxonomic hierarchy：分類の階層構造
taxonomic information：分類学（的）情報
taxonomic rank：分類学的階級、分類階級、分類順位
taxonomic ratio：分類比
taxonomic study：分類学的研究
taxonomic taxon：分類学的タクソン
taxonomical status：分類学的位置、分類学上の位置
taxonomy：分類法、分類学、分類

tear：引き裂く、引きちぎる
tear-shaped：ナスビ形の
technical assistance：技術支援
Teg：翅基片（Tegula）
tegula：（複 . -ae）、肩板（けんばん）、翅基片〔ラテン語〕
tegular arm：肩腕
tegumen (tg)：テグメン、覆片
tell apart：識別する、区別する
temperate climate：温帯気候
temperate forest：温帯林
temperate forest and relictual genus of Asian origin：アジア温帯林遺存属
temperate forest and wide-distributed genus：温帯森林広分布属
temperate forest and wide-distributed genus of American origin：アメリカ温帯林広分布属
temperate forest and wide-distributed genus of Asian origin：アジア温帯林広分布属
temperate genera：温帯属
temperate group：温帯グループ
temperate latitude：温帯緯度、温帯地方
temperate pierid butterfly：温帯（性）のシロチョウ科の蝶
temperate population：温帯域個体群
temperate rainforest：温帯雨林
temperate region：温帯地域
temperate species：温帯性の種
temperate zone：温帯
temperature：温度
temperature compensation：温度補償（性）
temperature condition：温度条件
temperature cycle：温度周期
temperature dependence：温度依存性

temperature effect：温度効果、温度の影響
temperature environment：温度環境
temperature fall：気温低下
temperature law：温度法則
temperature sensitivity：温度感受性
template：鋳型、テンプレート
template absorbance spectrum：吸収スペクトルテンプレート
temporal autocorrelation：時間的自己相関
temporal isolation：時間的隔離、一時的隔離
temporal scale：時間スケール、時間尺度
tendon cell：腱細胞
teneral：テネラル、不整成虫（羽化したてでまだ体がやわらかい成虫のこと）
tentacle：触角、触手、伸縮突起
tentacle nectary organ：伸縮突起様蜜腺、伸縮蜜腺
tentacle organ：伸縮突起
tentacular organ：触手状器官
tentatively：一応
teratological specimen：奇形標本
tergite：背板
tergum：背板
termen：外縁
terminal：末端
terminal branch：末端分岐
terminal fusion：末端融合型
terminal mechanism：末端機構
terminal spur：末端突起、末端の距
terminal stage of larva：幼虫末期
terpenoid defensive secretion：テルペノイド防御分泌物
terrestrial arthropod：陸生節足動物、陸域節足動物、陸産節足動物
terrestrial biome：地球上の生物群系、陸

のバイオーム
terrestrial crustacean：陸域甲殻類、陸産甲殻類、陸上甲殻類
terrestrial ecosystem：陸域生態系、陸上生態系
terrestrial habitat：地上の生息地
terrestrial life：地球生物、陸上生物
territorial behavior：占有行動、テリトリー行動、ナワバリ行動
territorial defence：ナワバリの防衛
territorial fluctuation：テリトリーの変動
territorial male：テリトリー占有雄
territorial marker：テリトリーマーカー
territorialism：ナワバリ制
territoriality：ナワバリ制
territorium：ナワバリ〔ラテン語〕
territory：テリトリー、縄張り、ナワバリ
territory contact area：テリトリー接触域
Tertiary period：第三紀（地質時代の新生代の一時代）
test of historical biogeographical hypothesis：歴史的生物地理学的仮説の検定
testis：（複．-es）、精巣〔ラテン語〕
Tethyan region：テチス区
Tethyan type：テチス型
tetracycline：テトラサイクリン
tetracycline hydrochloride：テトラサイクリンヒドロクロリド、テトラサイクリン塩酸塩
tetracycline-containing artificial diet：テトラサイクリン入り人工飼料
tetracycline-supplemented diet：テトラサイクリン補給飼料
tetrapyrrolic pigment：テトラピロール系色素

tga：肩腕（tegular arm）
thaw：解凍する
theoretical ground：理論的基礎、理論的根拠
theoretical model：理論的モデル
theoretical progress：理論的進展、理論的展開
theoretical treatment：理論的取り扱い
theory of island biogeography：島の生物地理学理論
theory of speciation：種分化（の）理論
thermal constant：有効積算温量定数、有効積算温度定数
thermal hysteresis protein：不凍タンパク質
thermo sensor：温度計、温度センサー
thermoperiod：温度周期
thermoperiodic response：温度周期反応
thermophase：高温期、温度相
thermoreceptor：温度受容器
thermoregulation：体温調節
thermoregulatory function：体温調節機能
thick ridge of cuticle：表皮の厚みのある隆起、クチクラの厚みのある隆起
thin membrane：薄膜
thinning the canopy：林冠の間引き
thoracic flight muscle：胸部の飛翔筋
thoracic leg：胸脚、胸部の脚
thoracic muscle：胸部筋肉（組織）
thoracic nervous system：胸部神経系
thoracic temperature：胸部温度
thorax：胸部（きょうぶ）〔ラテン語〕
thorax garter：胸部の帯糸
thorax mass：胸部重量
thorax shape：胸部の形
thorax width：胸部幅

thorn：とげ

thorough system of quality assurance：一貫した品質保証体制

threat：脅威

threaten：脅かす

threatened species：絶滅危惧種

threatened wildlife group：絶滅の恐れのある野生生物群、絶滅危惧の野生生物群

three generations a year：年三世代

three-allele system：三つの対立遺伝子システム

threshold：閾(いき)値

threshold intensity：閾(いき)値強度、閾値の強さ

threshold level：閾(いき)値

thrive：繁殖する、繁栄する

throughout the year：年中

Thysanoptera：アザミウマ目、総翅目

Thysanura：シミ目、総尾目

tibia：(複．-ae)、脛節(けいせつ)〔ラテン語〕

tibial：けい(脛)側

tibial spur：距刺、脛節棘

tibial tuft：脛節の毛房

tide：風潮、流れ

tiger moth：ヒトリガ

tiger pierid butterfly：ベニオビコバネシロチョウ

tiger swallowtail butterfly：メスグロトラフアゲハ

tiger-patterned morph：トラ模様のモルフォチョウ

tight genetic association：緊密な遺伝相関

tightly linked genetically：遺伝的に強く連鎖している

tightly linked locus：密に連鎖した遺伝子座

Tillyard notation：ティリヤード表記法

Tillyard's notation：ティリヤード表記法

Tillyardian notation：ティリヤード表記法

timber：材木

time measuring mechanism：測時機構

time of adult emergence：羽化時期

time of appearance：活動時間、出現時刻、姿を見せる時刻

time of disappearance：終了時刻、姿を消す時刻

time scale：時間スケール、時間尺度

time signal：時刻信号

time-measurement function：測時機能

time-measuring response：測時反応

time-period：時期

time-series data：時系列データ

time-since-colonisation：定着からの経過時間、移入からの経過時間

timing of pupation：蛹化時期

tip：先端、末端、崩す

tip of abdomen：尾端

tissue：組織(動植物の細胞の)

tissue paper：ティッシュペーパー

tissue tropism：組織向性、組織親和性

tissue-specific quantitative PCR technique：組織特有の定量的 PCR 法、組織特異的定量 PCR 法

titer：タイター、力価

title：表題

token stimulus：誘導する要因

tolerance：耐性

tolerance to cold：耐寒性

Toluidine blue：トルイジンブルー

tongue：口吻(こうふん)

tonic：緊張性の、持続性の

tooth-like：歯形の
toothed：ぎざぎざの
topographical：地形的
topographical model：地形図モデル
topography：地形
topotype：トポタイプ、原地基準標本、原地模式標本
torchlight：たいまつの火、懐中電灯の明り
tormogen：窩生（かせい）
tormogen cell：ソケット細胞、窩生細胞
tornal：後角部、肛角部、後縁角、肛角
tornus：後角部、肛角部、後縁角、肛角（「円」や「盤」を意味するラテン語）
torpor：休眠状態、冬眠状態
torrid zone：熱帯
torsion：ねじれ
total dry mass：総乾重量、総乾燥重量
total effective temperature：有効積算温度、有効温量、有効積算温量
total evidence analysis：全証拠解析
total map length：地図の全長
total mass：総重量
total range：総行動範囲
totipotency：全能性（ワーカーの）
touch：触覚
tough leaf：固い葉
toughening of leaf：葉の硬化
tourism：観光、観光事業
toxin：毒素、毒物
trabecula：柱、小柱、トラベキュラ
trachea：（複．-ae)、気管（きかん）〔ラテン語〕
trachea system：気管系
tracheal air sac：気管の気のう
tracheal system：気管系
tracheolar layer：毛細気管層
tracheole：気管小枝

tractable model：扱いやすいモデル
trade-off：トレードオフ、拮抗的関係
trading partner：貿易相手
trail marking：道しるべ
trail pheromone：道しるべフェロモン
trait：形質
trait-mediated indirect effect：形質の変化を介する間接効果
trajectory of evolution：進化軌跡、進化軌道
trampling：踏みつける
trans-：越えて -、他の側へ -、別の状態へ -〔ラテン語〕
trans-membrane receptor：トランスメンブラン受容体、膜貫通受容体
trans-migration：通過移動
transaminase：アミノ基転移酵素
transcription factor：転写制御因子、転写因子
transcriptional regulator：転写制御因子
transduce：変換する
transect：観察路、観察地、トランセクト
transect method：トランセクト法
transect recorder：トランセクトの記録者
transfer：媒介する、移動、注入する
transfer protein：運搬タンパク質
transform：変態する
transformation：変態、形質転換
transformation mechanism：変態機構
transformer：変化させるもの
transgenerational induction of defence：世代をわたる防御誘導、世代を越える防御誘導
transgenic pollen：トランスジェニック花粉、遺伝子導入花粉、形質転換花粉
transgression：違法、違反、犯罪

transient rise in temperature：過渡的な温度上昇
transiently：過渡的に
transit flight：通過移動飛翔
transition：転移型置換、転位、塩基転位
transition from race to a species：亜種から種への転移
transitional form：移行型
transitional zone of food plant：食草の移行帯
transliteration：換字
translocation：移動、移植、転地、転座
translucent：半透明の
transmembrane domain：膜貫通領域、膜貫通ドメイン
transparent bluish white marking：青白色の透明斑
transparent patch：透明な斑紋部分
transparent region：透過域
transpiration：蒸散
transplant：移植する
transport protein：運搬タンパク質
transversal：横（おう）、横（の）
transverse：横縫合線、横径の、横行の
transverse band：横行眼状紋
transverse plane：横断面
transverse section：横断面
transverse surface：横断面
transversion：転換型置換、転換
trap：トラップ
trap lining：トラップを敷くこと、トラップ覆工、トラップ用道づくり、わな道
trapezoidal sound pulse：台形状の音波、台形波のパルス音
trapping：罠（わな）、トラップを仕掛ける
traumatic insemination：トラウマを被った受精
travel up：走る（神経が）
traveling time：探索時間
Tre group：Tre系統
treacle：糖蜜
treated normal matriline：処理された正常な母系群
treatise：論文、学術論文、論説
treatment：処理、トリートメント
treatment chamber：処理室、処理チェンバー
treatment temperature：処理温度
tree canopy：樹冠
tree diagram：樹形図
tree nymph butterfly：オオゴマダラ
tree sap：樹液
tree species：樹種
tree trunk：樹幹、木の幹
treefall gap：倒木ギャップ
treeline：森林限界、高木限界線
treetop：こずえ、木の頂上
trehalase：トレハラーゼ
trehalose：トレハロース
tri-：3-、三 -〔ギリシャ語、ラテン語〕
triangle envelope：三角紙
triangular-shaped：三角形の
tribal maintenance：種族維持
tribe：族（分類階級の「ぞく」）
tribe name：族名
trichogen：毛生、生毛、発毛、毛母
trichogen cell：毛母細胞、生毛細胞
trichoid sensillum：感覚毛、毛状感覚子
trichome：毛、毛状突起
Trichoptera：トビケラ目、毛翅目
trichromatic：三色性の
trichromatic color vision：三色型色覚

tricin：トリシン
trifurcate：三枝に分かれる、三叉に分かれる
trim：切り取る
trimodal distribution：三峰性分布
trinomen：三語名〔ラテン語〕
trinomina：三語名〔ラテン語〕
trinominal name：三語名
triose phosphate isomerase：トリオースリン酸イソメラーゼ
Tris-EDTA buffer：TEバッファー
tritrophic：三栄養
trivial movement：日常飛翔、日常行動
trivoltine zone：三化性地域
trochanter：転節（てんせつ）
Troidini：ジャコウアゲハ族
trophobiotic partner：栄養共生パートナー
tropical arthropod：熱帯産節足動物
tropical arthropod dataset：熱帯産節足動物データセット
tropical beetle：熱帯産甲虫、熱帯性甲虫
Tropical Brushfoots：カバタテハ亜科
tropical butterfly：熱帯産の蝶、熱帯性の蝶、熱帯蝶
tropical climate：熱帯気候
tropical crepuscular butterfly：熱帯産の薄明活動性蝶
tropical dry：乾燥熱帯性
tropical dry vegetation：乾燥熱帯の植生
tropical forest：熱帯林
tropical invertebrate community：熱帯産無脊椎動物群集
tropical monsoon：熱帯モンスーン
tropical monsoon region：熱帯モンスーン地域
tropical morph：熱帯型
tropical old-growth forest：熱帯の老齢林
tropical rainforest：熱帯雨林、熱帯降雨林
tropical region：熱帯地方
tropical relic：熱帯遺存
tropical second-growth forest：熱帯の二次植生林
tropical seedbank dataset：熱帯種子銀行データセット
tropical trees：熱帯林
tropical zone：熱帯
tropics：熱帯地方
true alpine butterfly：純高山性蝶、真性高山蝶
True Brushfoots：タテハチョウ亜科
true butterflies, the (superfamily Papilionoidea)：アゲハチョウ上科
true diapause：真正休眠
true leg：真脚、真足
true phylogeography：真の系統地理学
true species richness：真の種数
true tropical species：純熱帯性の種
tube：管鞘、管
tubercle：伸縮突起、結節
tubular gland：管状腺
tuft：房、房状
Tukey least significant difference comparison：チューキー型最小有意差法の比較、チューキー型LSD法の比較
Tukey test：チューキー検定、チューキー法
tundra：ツンドラ
tundra environment：ツンドラ的環境
turf：芝生、芝土
turnover rate of population：個体群の入れ替わり率、個体群の回転率

turquoise：青緑色の
twice as likely：たぶん二倍
twig：小枝
twig mimic：小枝に擬態
twilight：日暮
twilight signal：日暮信号
two brood：二化
two-species stable equilibrium：二種の間での安定平衡
two-tailed Z-test：両側 Z 検定
two-way absorbance spectrum：双方向吸収スペクトル
TYLCV：トマト黄化葉巻ウイルス、トマト黄化葉巻病（Tomato Yellow Leaf Curl Virus）
tympanal chamber：鼓膜室、膜チャンバー
tympanal ear：鼓膜耳、鼓膜を持つ耳
tympanal membrane：鼓膜
tympanal organ：鼓膜器官
tympanal surface：鼓膜表面
tympanum：（複.-na）、鼓膜〔ラテン語〕
type：タイプ
type fixation：タイプ固定
type genus：タイプ属
type horizon：タイプ層準
type host：タイプ宿主
type locality：タイプ産地
type of distribution：分布系統、分布型
type series：タイプシリーズ
type species：タイプ種、模式種
type specimen：タイプ標本、模式標本、基準標本
typical autumn form：典型的な秋型
typification：タイプ化
tyrosine：チロシン

u

ubiquitous：至る所にある、偏在する
Ubx：超双胸遺伝子（*Ultrabithorax* gene）
ulnar：尺側の、尺骨（側）の
ultimate factor：究極要因
ultrabithorax gene：超双胸遺伝子
ultrasonic click：超音波クリック音
ultrasonic echolocation call：超音波の反響定位音、超音波の反響定位鳴声
ultrasound cry of bat：コウモリの超音波音の鳴き声
ultrasound-sensitive：超音波音感受性の
ultrasound-sensitive ear：超音波音感受性の耳
ultraviolet：紫外線
ultraviolet ray：紫外線
ultraviolet ray reflection：紫外線反射
ultraviolet region：紫外域
Ulysses butterfly：オオルリアゲハ
un-：でない -、反対に -〔英語〕
unaffect：影響を受けない
unaltered white：白色のまま、無変更の白色
unaltered yellow：黄色のまま、無変更の黄色
unavailability：不適格性
unavailable name：不適格名
unavailable nomenclatural act：不適格な命名法的行為
unavailable work：不適格な著作物
unbiased way：偏りのない方法
unblemished：きずのない
unclear：正体が分からない
uncommon species：珍しい種
uncus (un)：（交尾器の）ウンクス、鉤、交

尾器〔ラテン語〕
under-represent：小さな比率を占める
underbrush：下草
undergrowth：下草
underlying adaptive difference：根本的な適応の差、根本的な適応的相違
underlying membrane：基底膜
underlying population trend：個体数減少傾向
underneath：下面
underpin：強化する
underscore：強調する、力説する
underside (uns)：裏面
understory：低木層、下層
underway：進行中の
undeveloped ovariole：未発育の卵巣
undisturbed control：無攪乱の対照
undisturbed environment：攪乱されていない環境、未攪乱環境
undulate：起伏
UNEP：国連環境計画（United Nations Environment Programme）
unequal variance：不等分散
uneven crowding：不均一な混み具合
uneven sampling：一様でないサンプリング
unexpected mechanism：予想されていなかった機構、予期されなかった機構
UnF：前翅裏面（Underside of Forewing）
unfavorable habitat：不良な生息地、好ましくない生息地
unfertilized egg：未受精卵
unfertilized haploid egg：未受精の半数体卵
UnFH：両翅裏面（Underside of Forewing and Hindwing）

unfit：適応しない
unguis：（複.-ues）、爪〔ラテン語〕
UnH：後翅裏面（Underside of Hindwing）
uni-：1-、単-〔ラテン語〕
uniform dryness：均一な乾燥
uniform environment：均一な環境
uniform species density：一様な種密度
uniformity：斉一化（せいいつか）、斉一性
unifying framework：統一的な枠組み
unilateral pursuit：一方的な追跡、片側追跡
unimproved grassland：未改良の草地
uninfected female：非感染雌
uninominal：一語名の
uninominal name：一語名
unintentional introduction：非意図的導入、意図しない移入
uninvestigated：調査されていない
Union Jack butterfly：ベニヘリカザリシロチョウ
unique nymphalid species：特異なタテハチョウ種
unisexual sterility：単性不妊性、単性不稔性
unit of replication：複製単位
United Nations Environment Programme：国連環境計画
universal：普遍的
universal occurrence：普遍的な発生
universal pattern：ユニバーサルパターン、普遍的パターン
univoltine：一化性（いっかせい）
univoltine group：一化性集団
univoltine population：一化性個体群
univoltine type of egg overwintering：卵越冬一化型

unjustified emendation：不当な修正名
unknown：未知の
unlinked color pattern locus：非連鎖カラーパターン遺伝子座
unlinked region：未結合領域
unlogged forest：伐採されなかった森
unmanipulated control：無操作の対照
unobservable：観察不可能
unoccupied site：非占有場所
unpalatability：不快性
unpalatable：まずい、好みに合わない
unprecedented：先例のない
unprecedented decrease：空前の減少
unpredictable：予測不能な
unpredictable environment：予測不能な環境
unpredictable environmental exigency：予見不可能な環境急変
unpublished：未発表、未出版
unpublished work：未発表の著作物
unravel：解明する
unrelated species：無縁種
unreplicated mass sample：重複なしの大量サンプル
unrestrained collecting：自制のない採集
unseasonable development：季節はずれの発育
unsettled summer weather：不安定な夏の天候
untimely development：時期はずれの発育
untreated feminized matriline：無処理の雌化した母系
untreated insect：無処理の昆虫
unwary：油断、不注意
up-regulate：上方制御する
UpF：前翅表面（Upperside of Forewing）

UpFH：両翅表面（Upperside of Forewing and Hindwing）
UpH：後翅表面（Upperside of Hindwing）
uphold：維持する
upland：高地
uplift：隆起
upper lamina：翅表面
upper part of warm-temperate forest：暖帯林上部
upper surface：表翅、表面
upper threshold temperature：上方臨界温度
upperside (ups)：表面
upward selection：上方選択
Urania moth：ニシキオオツバメガ、ツバメガ科オオツバメガ亜科の蛾
urban area：都市（的）地域
urge：推進する
uric acid：尿酸
uric acid derivative：尿酸誘導体
urosome：腹部
used species：利用種
Ussuri type：ウスリー型
utopian：夢物語（的な）、理想郷の
UTR：非翻訳領域（UnTranslated Region）
UV：紫外線（UltraViolet）

v

V：裏面、下面、腹側（Ventral）
V-shaped marking：V字形の斑紋
vacant patch：空きパッチ
vacant site：空白地、空き地
vacant spot：空白地帯
vacuum：真空（状態）
valid name：有効名

valid nomenclatural act：有効な命名法的行為
validated：有効にされた
validity：有効性
valva：（複 .-ae）、（交尾器の）バルバ、弁〔ラテン語〕
van't Hoff's law：ファントホッフの法則
van't Hoff's principle：ファントホッフの法則
var.：変種（varietas の略語）、「『属名 種名 var. 品種名』の形式で表記」
variability：変異性、変動性、多様性
variable phenotype：可変表現型
variance：分散
variant：変異体、変動体、多様体、変異集団
variant population：変異個体群
variant spelling：変体綴(つづ)り
variation：変異、変動
variation pattern：変異様式
variation series：変異系列
variation speed：変異速度
variety：変種、品種、系統、多様性
vary：変える、変化する
vas deferens：輸精管
vascular plant：維管束植物
vector：媒介者、ベクター、媒介する
vector control：媒介者防除
vector of disease：病原菌媒介生物、疾病媒介生物
vector-borne disease：媒介者によって媒介される疾病、ベクター媒介性疾患
vegetation：植生、植物
vegetation zone：植生帯
vegetative growth：栄養成長
vegetative reproduction：栄養繁殖
vegetative requirement：栄養要求
vein：翅脈、葉脈、脈
vein pattern：翅脈パターン
vein-cutting behavior：葉脈切断行動
venation：脈相、脈系、翅脈
ventilation：換気、通気性
ventral：腹部の、腹側の、裏面の、腹側（ふくそく）
ventral hindwing：腹側の後翅
ventral light：腹側からの光
ventral retina：腹側網膜
ventral side：腹側
ventral surface：腹側表面
ventral view：腹側から見た図
ventral wing marking：腹側の翅の斑紋
ventral wing surface：腹側の翅面、翅の裏面
verge of extinction：絶滅の瀬戸際
vermilion：朱色
Vermont stream：バーモント州の河川
vernacular name：俗名、通俗名
vertebrate：脊椎動物
vertebrate predator：脊椎動物の捕食者
vertebrate retina：脊椎動物網膜
vertex：頭頂部(とうちょうぶ)
vertical distribution：垂直分布
vertical infection：垂直感染
vertical stratification：垂直層別
vesica (ve)：ベシカ、のう、内鞘
vestibulum：腔前室
vestigial mutualistic benefit：退化した相利共生的利益
vestigial organ：痕跡器官、退化した器官
viability：生存率、生存度、生存力、存続性
viability selection：生存力選択

viable：生存力のある

viable offspring：生存可能な子孫

viable population：存続可能個体群

vibration property：振動特性

vibrational property：振動特性

vibratory papilla：振動突起

vicariance：分断分布

vice versa, and：逆もまた同様、逆に

Viceroy butterfly：チャイロイチモンジ、カバイロイチモンジ

vicious circle：悪いサイクル、悪循環

Victoria era：ビクトリア朝

vinculum (vin)：繋帯、ヴィンクルム、ビンクルム

vineyard：ぶどう園

virtual tautonymy：疑似同語反復

virulence：病毒性、病原力

virus：(複. -es)、ウイルス〔ラテン語〕

viscous：粘性のある

visible light spectrum：可視光スペクトル

visual cell：視細胞

visual communication：視覚による通信、視覚通信

visual cue：視覚刺激

visual inspection：目視検査

visual organ：視覚器官

visual pigment：視物質

visual sense：視覚

visual signal：視覚信号

visual stimulus：(複. -li)、視覚刺激〔ラテン語〕

visual system：視覚系

vitamin：ビタミン

vitellin：ビテリン

vitelline membrane：卵黄膜

vitellogenesis：卵黄形成

vitellogenin：ビテロジェニン

vivid color pattern：鮮やかなカラーパターン

viviparous：胎生の

vocalization of predatory bird：捕食鳥のさえずり、捕食鳥の鳴き声

Vogel's organ：フォーゲル器官

volatile compound：揮発性化合物

volatile ketone：揮発性ケトン

volcanic：火山性の

volcanic activity：火山活動

volcanic island：火山島

voltinism：化性（かせい）

voucher：証拠物件

voucher specimen：証拠標本

vulnerable：危急

vulnerable body：傷つきやすい身体、傷つきやすい体

vulnerable site：傷つきやすい場所

vulnerable species：危急種

w

waiting type：待ち伏せ型

Wald statistic：ワルド統計量

walking：歩行

wall pattern：壁模様

Wallace's line：ウォーレス線

wandering：歩き回る

wandering behavior：徘徊行動

war of attrition：消耗戦

warm blooded：定温の、温血の

warm-season species：暖地性種

warm-temperate ancestral type：暖温帯性の祖先型

warm-temperate forest and wide-distributed genus of Asian origin：アジ

ア暖帯林広分布属

warm-temperate species：暖地性種、暖温帯の種
warmer tropical rainforest：熱帯雨林
warming：温暖化
warning color：警戒色
warning color pattern：警告色パターン
warning coloration：警告色、警戒色
warning pattern：警告パターン
warning signalling：警告信号
warningly colored and mimetic butterfly：警告色化と擬態化した蝶
warningly colored butterfly：警告色の蝶
Washington Convention：ワシントン条約
wasp：黄蜂、ハチ、スズメバチ
waste product：老廃物
watching：観察
water purification：水質浄化
water strider：アメンボ
wavelength region：波長領域
wavelength specific behavior：波長特異的な行動
wavy line：波状の線
WCMC：世界自然保全モニタリングセンター（World Conservation Monitoring Centre）
weakened state：弱体化した状態
weakly developed eyespot：発達の悪い目玉模様
weather condition：気象条件
web：クモの巣
Weber's line：ウェーバー線
wedge-shaped：くさび形の
weed：雑草
weight loss：体重の減少（率）
weight of post-diapause larva：休眠後の幼虫の体重、後休眠期の幼虫の体重
well-understood molecular mechanism：よく理解された分子機構
West-Chinese element：中国西部系要素
western pygmy blue butterfly：アレチコビトシジミ（アレチコシジミ：改訂新名称）
wet forest：湿潤森林
wet lowland grass：低湿地の草、湿地草
wet season：雨季
wet season form：雨季型
wet-dry seasonal environment：乾湿の季節的環境
wetland：湿地
wg：無翅遺伝子（*wingless* gene）
wheat field：小麦畑
whip-like：鞭状の
white band：白色の帯
white forewing and hindwing shutter allele：白色の前翅と後翅のシャッター対立遺伝子
white forewing portion colored yellow：前翅の白色部分を黄色に変えた
white line：白い線
white list：ホワイトリスト
white marking：白色の斑紋
white morph：白化型
white morph formation：白化型形成
white morph induction：白化型誘起
white morph production：白化型生産
white patch：白い大斑点
white pupation board：白色の蛹化台紙
white scale：白色鱗粉
white streak：白いすじ
white/yellow color shift：白色／黄色の色彩転換

white/yellow switch：白色／黄色スイッチ
Whites：シロチョウ亜科、シロチョウ科
Whites and Sulphurs：シロチョウ科
whitish：白っぽい、やや白い
whole mount：全載標本、ホールマウント、全組織標本、伸展標本
whole view：全体像
whole-genome bacterial artificial chromosomal library：全ゲノムバクテリア人工染色体ライブラリー
wide distribution：広分布、広域分布
wide variation ability：広い変異能力
wide-distributed species：広分布種、広域分布種
widely distributed genera：広分布の属、広分布属
widely distributed species：広分布種、広域分布種
widening ride：乗馬道の拡大
wider countryside species：広範な里山にいる種
widespread coexistence：広範囲にわたる共存
widespread extinction：広範囲に及ぶ絶滅
widespread species：広く分布する種
width of membrane：膜（の）幅
width of outer membrane：外膜（の）幅
wild adult：野外成虫
wild *Drosophila*：野生のショウジョウバエ
wild flower：野草、野生の花
wild individual：野生個体
wild population：野生個体群

wild, in the：野外では
wild-caught individual：野外採集個体
wild-caught male：野外採集雄、野外で採集した雄
wildlife sanctuary：自然保護区
wind：風
wind-borne movement：風まかせ移動
wind-tunnel assay：風洞実験法
window：窓
window screen：窓（の）網戸
wing：翅（はね）
wing area：翅面積
wing base：翅基部
wing base nerve branch：翅基部の神経枝
wing beat：羽ばたき
wing bud：翅芽
wing cell：翅細胞
wing color preference cue：翅色選好の刺激
wing condition：翅の損傷度、翅の状態
wing configuration：翅型
wing dimorphism：翅二型
wing disc：翅原基
wing edge：翅端
wing epidermis：翅の表皮細胞
wing flick：翅による摩擦音、翅によるフリック音
wing length：翅長
wing load：翼（翅）荷重
wing loading：翅荷重、翅面荷重
wing margin：翅の外縁部
wing morphology：翅形態
wing pattern diversification：翅パターンの多様性
wing pattern polyphenism：翅パターンの表現型多型

wing patterning candidate gene：翅のパターン形成候補遺伝子
wing polymorphism：翅多型、翅型多型
wing portion：翅の部位
wing scale：翅の鱗粉
wing sharp：翅形
wing span：翅長、開張
wing subdivision：亜翅室
wing surface：翅（表）面
wing tissue：翅組織
wing toughness：翅の強度、翅の硬さ
wing upperside：翅の表面
wing variation：翅変異
wing vein：翅脈（しみゃく）
wing vein pattern：脈相
wing venation：翅の脈相、翅の脈系
wing-coupling mechanism：連結器官
wing-pigment：翅の色素
wing-related trait：翅関連形質
wing-tip：翅の先端
wingbeat：はばたく
winged form：有翅型、翅型
wingless form：無翅型
wingless gene：無翅遺伝子
wingspan：開張、翅幅
winter bud：冬芽
winter cold：冬期の寒さ
winter day：冬日
winter diapause：冬眠、冬休眠
winter survival：越冬の成功率、冬期の生存率、冬の生存率
winter temperature：冬期の気温、冬季気温
wintering colony：越冬集団繁殖地
winterkill：冬枯
wire mesh enclosure：金網で囲まれた入れ物
withdrawal：回収、撤回
within-morph courtship：同色の翅内の求愛
within-network variation：ネットワーク内変異
without irradiation：暗条件で、無照射で
withstand：耐える
Wolbachia：ウォルバキア、ボルバキア
Wolbachia density：ボルバキア密度
Wolbachia-induced feminization：ボルバキアによる雌化
Wolbachia-induced male killing：ボルバキアによる雄殺し
Wolbachia-induced parthenogenesis：ボルバキアによる単為生殖
Wolbachia-induced reproductive manipulation：ボルバキアによる生殖操作
Wolbachia-mediated addiction：ボルバキア媒介による中毒
Wolbachia-mediated addiction hypothesis：ボルバキア媒介による中毒説
Wolbachia-mediated addiction mechanism：ボルバキア媒介による中毒機構
wood nymph butterfly：ペガラオオモンヒカゲ（暫定和名）（ジャノメチョウ科の一種）
Wood Nymphs：ジャノメチョウ科
wood pasture：森林放牧地
woodfuel：木質燃料、木材燃料
woodland：森林地帯、森林地、森林性の
woodland butterfly：森林性の蝶
woodland environment：森林的環境
woodland habitat：森林性生息場所
woodland species：森林性種

woodlouse：ワラジムシ
woody plant parasite：木本寄生
woolly bear：毛虫
work：著作物
work of an animal：動物の仕業
worker policing：ワーカーポリシング
working hypothesis：作業仮説
World Conservation Monitoring Centre：世界自然保全モニタリングセンター
world's famous butterfly：世界の名蝶
worthwhile：やりがいのある、費やした時間に相当する
wound：傷
wound-induced defence：傷害で誘導される防御、傷害誘導防御
wriggling movement：蠕動運動（くねらせて出てくること）
wsp gene segment：wsp 遺伝子断片（Wolbachia surface protein）

x

xanthopterin：キサントプテリン

y

y-intercept：Y 切片
year-to-year variation：年変動
yellow band：イエローバンド
yellow pigment：黄色色素
yellow stripes colored white：黄色の縞模様を白色に変えた
yellow-banded female model：黄色翅の雌モデル
yellow-green pigment：黄緑色素
yellowing syndorome：イエローイングシンドローム、黄変症候群
young caterpillar：若齢幼虫
young fruit：未熟果
young pupa：蛹初期
Yucca giant skipper butterfly：イトランセセリ

z

Z chromosome：Z 染色体
Z-linked single copy nuclear locus：Z 染色体と連鎖した単コピー核遺伝子座
zebra longwing butterfly：キジマドクチョウ
zebra swallowtail butterfly：トラフタイマイ
zen gene：ゼン遺伝子、ツェアクヌルト（*zerknullt*）遺伝子（キイロショウジョウバエのホメオティック遺伝子）〔ドイツ語〕
Zephyrus：ゼフィルス
zero growth isocline：ゼロ成長の等傾斜線
zigzag pattern：ジグザグ模様
zodiac moth：オジロルリツバメガ、フトオビルリツバメガ
zonation：帯状分布
zone (zn)：ゾーン
zoocarotenoid：動物性カロチノイド
zoogeographical region：動物地理区
zoological formula：動物定型名
zoological name：動物学の名称
zoological nomenclature：動物命名法
zoological taxon：動物学的タクソン
zoologist：動物学者
Zoraptera：ジュズヒゲムシ目、絶翅目
zygote：接合子
zygote mortality：接合子の死亡率

付録：日本産蝶類名称の英和／和英編

付録：日本産蝶類名称の英和／和英編

■ 英和編

A

African Grass Blue：ハマヤマトシジミ
Albocaerulean：サツマシジミ
Alphabetical Hairstreak：ウラミスジシジミ
Alpine Clouded Yellow：
　ミヤマモンキチョウ
Ambigua Fritillary：コヒョウモンモドキ
Anadyomene Fritillary：クモガタヒョウモン
Angled Castor：カバタテハ
Angled Sunbeam：ウラギンシジミ
Angulated Grass Yellow：ツマグロキチョウ
Apefly：シロモンクロシジミ
Arctic Skipper：タカネキマダラセセリ
Argus Rings：ヒメウラナミジャノメ
Argyrognomon Blue：ミヤマシジミ
Arran Brown：クモマベニヒカゲ
Asahina's Skipper：アサヒナキマダラセセリ
Asama Silver-studded Blue：アサマシジミ
Asama White Admiral：アサマイチモンジ
Asamana Arctic：タカネヒカゲ
Asian Comma：キタテハ
Asian Swallowtail：アゲハ
Autumn Leaf：イワサキコノハ

B

Bamboo Treebrown：シロオビヒカゲ
Banana Skipper：バナナセセリ
Bianor Peacock：カラスアゲハ
Black Cupid：クロツバメシジミ
Black Hairstreak：リンゴシジミ
Black Veined Tiger：スジグロシロマダラ
Black-banded Hairstreak：
　ミズイロオナガシジミ
Black-veined White：エゾシロチョウ
Blackburn's Blue：サツマシジミ
Blue Admiral：ルリタテハ
Blue Branded King Crow：
　マルバネルリマダラ
Blue Hairstreak：ウラゴマダラシジミ
Blue Pansy：アオタテハモドキ
Blue Quaker：ツシマウラボシシジミ
Blue Spotted Crow：ミダムスルリマダラ
Blue Tiger：ミナミコモンマダラ
Blue-branded King Crow：
　マルバネルリマダラ
Blue-spotted Crow：ミダムスルリマダラ
Brilliant Hairstreak：アイノミドリシジミ
Brimstone：ヤマキチョウ
Broad-bordered Grass Yellow：
　ホシボシキチョウ
Brown Awl：タイワンアオバセセリ
Brown-banded Hairstreak：
　ウスイロオナガシジミ

C

Cabbage White：モンシロチョウ
Camberwell Beauty：キベリタテハ
Cassia Butterfly：ウスキシロチョウ
Ceylon Blue Glassy Tiger：
　リュウキュウアサギマダラ
Ceylon Lesser Albatross：ナミエシロチョウ
Checkered Skipper：タカネキマダラセセリ
Chequered Skipper：タカネキマダラセセリ
Chestnut Tiger：アサギマダラ
China Flat：ダイミョウセセリ
Chinese Bushbrown：ヒメジャノメ
Chinese Comma：キタテハ
Chinese Peacock：カラスアゲハ
Chinese Windmill：ジャコウアゲハ
Chinese Yellow Swallowtail：アゲハ
Chocolate Albatross：タイワンシロチョウ
Chocolate Argus：イワサキタテハモドキ
Chocolate Pansy：イワサキタテハモドキ
Citrus Swallowtail：オナシアゲハ
Cognatus Green Hairstreak：
　ジョウザンミドリシジミ
Comma：シータテハ
Common Albatross：ナミエシロチョウ
Common Awl：テツイロビロウドセセリ

Common Banded Awl :
　オキナワビロウドセセリ
Common Bluebottle : アオスジアゲハ
Common Crow Butterfly : ガランピマダラ
Common Eggfly : リュウキュウムラサキ
Common Evening Brown :
　ウスイロコノマチョウ
Common Five-ring : ヒメウラナミジャノメ
Common Grass Yellow :
　キチョウ（ミナミキチョウ）
Common Hedge Blue : ヤクシマルリシジミ
Common Indian Crow : ガランピマダラ
Common Jay : ミカドアゲハ
Common Leopard : ウラベニヒョウモン
Common Lineblue : ヒメウラナミシジミ
Common Map : イシガケチョウ
Common Mormon : シロオビアゲハ
Common Palm Dart : ネッタイアカセセリ
Common Rose : ベニモンアゲハ
Common Rose Swallowtail :
　ベニモンアゲハ
Common Sailer : コミスジ
Common Sergeant : シロミスジ
Common Tiger : スジグロカバマダラ
Common Yellow Swallowtail : キアゲハ
Compton Tortoiseshell : エルタテハ
Constable : スミナガシ
Copper Hairstreak : クロミドリシジミ
Cranberry Blue : カラフトルリシジミ
Crow Eggfly : ヤエヤマムラサキ
Cycad Blue Butterfly :
　クロマダラソテツシジミ
Cycad Butterfly *** : ソテツシジミ

D

Daisetsuzana Arctic : ダイセツタカネヒカゲ
Danaid Eggfly : メスアカムラサキ
Daphne Fritillary : ヒョウモンチョウ
Dark Cerulean : ルリウラナミシジミ
Dark Evening Brown : クロコノマチョウ
Dark Grass Blue : ハマヤマトシジミ
Dark Green Fritillary : ギンボシヒョウモン
Diamina Fritillary :
　ウスイロヒョウモンモドキ
Diana Treebrown : クロヒカゲ
Dingy Line Blue : マルバネウラナミシジミ
Dingy Lineblue : マルバネウラナミシジミ
Dod-dash Sailer : ホシミスジ
Double-branded Crow : ルリマダラ
Dragon Swallowtail : ホソオチョウ
Dryad : ジャノメチョウ
Dryad Butterfly : ジャノメチョウ

E

East European Sailer : フタスジチョウ
Eastern Large Blue : オオゴマシジミ
Eastern Pale Clouded Yellow : モンキチョウ
Eastern Silverstripe :
　ウラギンスジヒョウモン
Eastern Wood White : ヒメシロチョウ
Essex Skipper : カラフトセセリ
European Beak : テングチョウ
European Map Butterfly : アカマダラ
European Purple Emperor : コムラサキ
European Skipper : カラフトセセリ
Evening Brown : ウスイロコノマチョウ
Eversmann's Parnassius :
　キイロウスバアゲハ
Eyed Pansy : アオタテハモドキ

F

False Comma : エルタテハ
False Heath Fritillary :
　ウスイロヒョウモンモドキ
False Ringlet : ヒメヒカゲ
Flower Swift : オオチャバネセセリ
Forest Pierrot : ゴイシシジミ
Forest Quaker : リュウキュウウラボシシジミ
Forget-me-not :
　ムラサキオナガウラナミシジミ
Formosan Swift : ユウレイセセリ
Freija Fritillary : アサヒヒョウモン

付録：日本産蝶類名称の英和／和英編

Freya's Fritillary：アサヒヒョウモン
Freyer's Purple Emperor：コムラサキ
Fujisan Green Hairstreak：フジミドリシジミ

G

Glacial Apollo：ウスバアゲハ
Glacial Parnassius：ウスバアゲハ
Glassy Tiger：ヒメアサギマダラ
Golden Birdwing：キシタアゲハ
Golden Hairstreak：ウラキンシジミ
Goschkevitshi's Labyrinth：
　　サトキマダラヒカゲ
Gram Blue：オジロシジミ
Grass Demon：オオシロモンセセリ
Gray-pointed Pierrot：クロシジミ
Gray-veined White：スジグロシロチョウ
Great Eastern Silverstripe：
　　オオウラギンスジヒョウモン
Great Eggfly：リュウキュウムラサキ
Great Mormon：ナガサキアゲハ
Great Nawab：フタオチョウ
Great Orange Tip：ツマベニチョウ
Great Purple：オオムラサキ
Great Purple Emperor：オオムラサキ
Green Flash：イワカワシジミ
Green Hairstreak：ミドリシジミ
Green-veined White：
　　エゾスジグロシロチョウ
Green-veined White：
　　ヤマトスジグロシロチョウ
Grizzled Skipper：ヒメチャマダラセセリ

H

Hayashi Hairstreak：ハヤシミドリシジミ
Heath Fritillary：コヒョウモンモドキ
High Brown Fritillary：ウラギンヒョウモン
Hill Hedge Blue：ルリシジミ
Hisamatsu Green Hairstreak：
　　ヒサマツミドリシジミ
Holly Blue：ルリシジミ
Hungarian Glider：フタスジチョウ

Hylas Common Sailer：リュウキュウミスジ

I

Indian Awlking：アオバセセリ
Indian Cabbage White：
　　タイワンモンシロチョウ
Indian Cupid：タイワンツバメシジミ
Indian Fritillary：ツマグロヒョウモン
Indian Leaf Butterfly：コノハチョウ
Indian Palm Bob：クロボシセセリ
Indian Red Admiral：アカタテハ
Ino Fritillary：コヒョウモン
Iphigenia Fritillary：カラフトヒョウモン

J

Jacintha Eggfly：リュウキュウムラサキ
Janson's Swift：ミヤマチャバネセセリ
Japanese Argus：ベニヒカゲ
Japanese Circe：ゴマダラチョウ
Japanese Dart：キマダラセセリ
Japanese Emperor：オオムラサキ
Japanese Flash：トラフシジミ
Japanese Labyrinth：ヤマキマダラヒカゲ
Japanese Luehdorfia：ギフチョウ
Japanese Oakblue：ムラサキシジミ
Japanese Rings：ウラナミジャノメ
Japanese Scrub Hopper：
　　ホシチャバネセセリ
Japanese Silverlines：キマダラルリツバメ
Japanese Swift：コチャバネセセリ
Jezo Green Hairstreak：エゾミドリシジミ
Jonasi Orange Hairstreak：ムモンアカシジミ

K

Korean Hairstreak：チョウセンアカシジミ

L

Large Banded Swift：トガリチャバネセセリ
Large Brown：オオヒカゲ
Large Cabbage White：オオモンシロチョウ
Large Comma：エルタテハ

Large High Brown：
　オオウラギンヒョウモン
Large Map Butterfly：サカハチチョウ
Large Sailer：オオミスジ
Large Shijimi Blue：オオルリシジミ
Large Skipper：コキマダラセセリ
Large Tortoiseshell：ヒオドシチョウ
Large Tree Nymph：オオゴマダラ
Large White：オオモンシロチョウ
Latifasciatus Green Hairstreak：
　ヒロオビミドリシジミ
Lemon Butterfly：オナシアゲハ
Lemon Emigrant：ウスキシロチョウ
Lemon Migrant：ウスキシロチョウ
Leoninus Skipper：スジグロチャバネセセリ
Lesser Albatross：ナミエシロチョウ
Lesser Brimstone：スジボソヤマキチョウ
Lesser Grass Blue：シルビアシジミ
Lesser Marbled Fritillary：コヒョウモン
Lilacine Bushbrown：コジャノメ
Lime Butterfly：オナシアゲハ
Lime Swallowtail：オナシアゲハ
Long Tail Spangle：オナガアゲハ
Long-streak Sailer：ミスジチョウ
Long-tailed Blue：ウラナミシジミ
Lycormas Blue：カバイロシジミ

M

Maackii Peacock：ミヤマカラスアゲハ
Maculatus Skipper：チャマダラセセリ
Malayan：タイワンクロボシシジミ
Malayan Crow：シロオビマダラ
Malayan Eggfly：ヤエヤマムラサキ
Mandarin Grass Yellow：キタキチョウ
Mangrove Tree Nymph：オオゴマダラ
Map Butterfly：アカマダラ
Marbled Fritillary：ヒョウモンチョウ
Marginalis Treebrown：クロヒカゲモドキ
Masaki's Rings：マサキウラナミジャノメ
Melissa Arctic：ダイセツタカネヒカゲ
Mera Black Hairstreak：ミヤマカラスシジミ

Metallic Cerulean：シロウラナミシジミ
Moore's Cupid：ゴイシツバメシジミ
Moorland Clouded Yellow：
　ミヤマモンキチョウ
Mottled Emigrant：ウラナミシロチョウ
Mountain Tortoiseshell：コヒオドシ
Mourning Cloak：キベリタテハ

N

Nettle-tree Butterfly：テングチョウ
Northern Checkered Skipper：
　カラフトタカネキマダラセセリ
Northern Orange Hairstreak：
　カシワアカシジミ（キタアカシジミ）

O

Ocean Tree-Nymph：オオゴマダラ
Ochracea Skipper：ヒメキマダラセセリ
Ogasawara Hedge Blue：オガサワラシジミ
Ogasawara Swift：オガサワラセセリ
Okinawa Peacock：オキナワカラスアゲハ
Old World Swallowtail：キアゲハ
Orange Emigrant：キタウスキシロチョウ
Orange Hairstreak：アカシジミ
Orange Oak Leaf：コノハチョウ
Orange Oakleaf：コノハチョウ
Orange Tiger：スジグロカバマダラ
Orange Tip：クモマツマキチョウ
Oriental Black-veined White：
　ミヤマシロチョウ
Oriental Blue Tiger：ウスコモンマダラ
Oriental Chequered Darter：アカセセリ
Oriental Hairstreak：オオミドリシジミ
Oriental Palm Bob：クロボシセセリ
Orion Blue：ジョウザンシジミ

P

Painted Lady：ヒメアカタテハ
Pale Clouded Yellow：モンキチョウ
Pale Grass Blue：ヤマトシジミ
Pale Hedge Blue：タッパンルリシジミ

付録：日本産蝶類名称の英和／和英編

Pale Palmdart : ネッタイアカセセリ
Pallas's Fritillary : ウラギンスジヒョウモン
Palm Bob : クロボシセセリ
Paper-butterfly : オオゴマダラ
Pea Blue : ウラナミシジミ
Peacock : クジャクチョウ
Peacock Pansy : タテハモドキ
Plain Tiger : カバマダラ
Plains Cupid : クロマダラソテツシジミ
Plumbago Blue : カクモンシジミ
Poplar Admiral : オオイチモンジ
Powdered Oakblue : ムラサキツバメ
Pseudo-Labyrinth : キマダラモドキ
Psyche : クロテンシロチョウ
Purple Beak : ムラサキテングチョウ

Q

Quaker : ヒメウラボシシジミ

R

Red Helen : モンキアゲハ
Red-ring Circe : アカボシゴマダラ
Red-spotted Hairstreak :
　ベニモンカラスシジミ
Restricted Demon : クロセセリ
Reverdin's Blue : ミヤマシジミ
Rice Paper Butterfly : オオゴマダラ
Rice Swift : ユウレイセセリ
Riukiuana Chinese Bushbrown :
　リュウキュウヒメジャノメ
Riukiuana Rings :
　リュウキュウウラナミジャノメ
Rustic : タイワンキマダラ

S

Sagana Fritillary : メスグロヒョウモン
Sailer : ミスジチョウ
Saphirinus Green Hairstreak :
　ウラジロミドリシジミ
Scarce Heath : シロオビヒメヒカゲ
Scarce Large Blue : ゴマシジミ

Scotosia Fritillary : ヒョウモンモドキ
Sericin Swallow-tail Butterfly :
　ホソオチョウ
Short-tailed Blue : ツバメシジミ
Sicelis Treebrown : ヒカゲチョウ
Silver Forget-me-not :
　ウスアオオナガウラナミシジミ
Silver Hairstreak : ウラクロシジミ
Silver-lined Skipper : ギンイチモンジセセリ
Silver-spotted Skipper : ホソバセセリ
Silver-studded Blue : ヒメシジミ
Silver-washed Fritillary : ミドリヒョウモン
Sivery Hedge Blue : ルリシジミ
Sky Blue : アサマシジミ
Small Branded Swift : チャバネセセリ
Small Cabbage White : モンシロチョウ
Small Copper : ベニシジミ
Small Grass Yellow : ホシボシキチョウ
Small Labyrinth : ヒメキマダラヒカゲ
Small Luehdorfia : ヒメギフチョウ
Small Straight Swift : ヒメイチモンジセセリ
Small Tortoiseshell : コヒオドシ
Small White : モンシロチョウ
Smaragdinus Green Hairstreak :
　メスアカミドリシジミ
Snout Butterfly : テングチョウ
Spangle : クロアゲハ
Spring Flat : ミヤマセセリ
Staff Sergeant : ヤエヤマイチモンジ
Strait Swift : イチモンジセセリ
Striped Blue Crow : ツマムラサキマダラ
Sugitani's Hedge Blue : スギタニルリシジミ
Sulphur : ヤマキチョウ
Swinhoe's Chocolate Tiger :
　タイワンアサギマダラ
Sylvaticus Skipper :
　ヘリグロチャバネセセリ

T

Tailed Cupid : タイワンツバメシジミ
Tailed Jay : コモンタイマイ

Tailless Bushblue : ルーミスシジミ
Tailless Hairstreak : コツバメ
Teleius Large Blue : ゴマシジミ
Thor's Fritillary : ホソバヒョウモン
Thore Fritillary : ホソバヒョウモン
Three Spots Grass Yellow : タイワンキチョウ
Three-spot Grass Yellow : タイワンキチョウ
Tiny Grass Blue : ホリイコシジミ
Transparent 6-Line Blue :
　　アマミウラナミシジミ
Transparent Six-line Blue :
　　アマミウラナミシジミ
Trebellius Flat : コウトウシロシタセセリ
Tree Nymph Butterfly : オオゴマダラ
Two-brand Crow : ルリマダラ

V

Varied Eggfly : リュウキュウムラサキ

W

Walnut Hairstreak : オナガシジミ
White Admiral : イチモンジチョウ
White Albatross : カワカミシロチョウ
White Banded Awl :
　　タイワンビロウドセセリ
White Clouded Parnassius :
　　ヒメウスバアゲハ
White Mountain Arctic :
　　ダイセツタカネヒカゲ
White-bordered : キベリタテハ
White-letter Hairstreak : カラスシジミ
White-tipped Woodland Brown :
　　ツマジロウラジャノメ
Wonderful Green Hairstreak :
　　キリシマミドリシジミ
Wood White : エゾヒメシロチョウ
Woodland Brown : ウラジャノメ

Y

Yayeyamana Peacock *** :
　　ヤエヤマカラスアゲハ

Yayeyamana Rings :
　　ヤエヤマウラナミジャノメ
Yellow Apollo : キイロウスバアゲハ
Yellow Awl : キバネセセリ
Yellow Tip : ツマキチョウ
Yellow-legged Tortoiseshell :
　　ヒオドシチョウ

Z

Zebra Blue : カクモンシジミ
Zebra Hairstreak : ウラナミアカシジミ
Zigzag Fritillary : アサヒヒョウモン

■ 和英編

ア

アイノミドリシジミ : Brilliant Hairstreak
アオスジアゲハ : Common Bluebottle
アオタテハモドキ : Blue Pansy
アオタテハモドキ : Eyed Pansy
アオバセセリ : Indian Awlking
アカシジミ : Orange Hairstreak
アカセセリ : Oriental Chequered Darter
アカタテハ : Indian Red Admiral
アカボシゴマダラ : Red-ring Circe
アカマダラ : European Map Butterfly
アカマダラ : Map Butterfly
アゲハ : Asian Swallowtail
アゲハ : Chinese Yellow Swallowtail
アサギマダラ : Chestnut Tiger
アサヒナキマダラセセリ :
　　Asahina's Skipper
アサヒヒョウモン : Freija Fritillary
アサヒヒョウモン : Freya's Fritillary
アサヒヒョウモン : Zigzag Fritillary
アサマイチモンジ : Asama White Admiral
アサマシジミ : Asama Silver-studded Blue
アサマシジミ : Sky Blue
アマミウラナミシジミ :
　　Transparent Six-line Blue

付録：日本産蝶類名称の英和／和英編

アマミウラナミシジミ：
　Transparent 6-Line Blue

イ

イシガケチョウ：Common Map
イチモンジセセリ：Strait Swift
イチモンジチョウ：White Admiral
イワカワシジミ：Green Flash
イワサキコノハ：Autumn Leaf
イワサキタテハモドキ：Chocolate Argus
イワサキタテハモドキ：Chocolate Pansy

ウ

ウスアオオナガウラナミシジミ：
　Silver Forget-me-not
ウスイロオナガシジミ：
　Brown-banded Hairstreak
ウスイロコノマチョウ：
　Common Evening Brown
ウスイロコノマチョウ：Evening Brown
ウスイロヒョウモンモドキ：
　Diamina Fritillary
ウスイロヒョウモンモドキ：
　False Heath Fritillary
ウスキシロチョウ：Cassia Butterfly
ウスキシロチョウ：Lemon Emigrant
ウスキシロチョウ：Lemon Migrant
ウスコモンマダラ：Oriental Blue Tiger
ウスバアゲハ：Glacial Apollo
ウスバアゲハ：Glacial Parnassius
ウラギンシジミ：Angled Sunbeam
ウラキンシジミ：Golden Hairstreak
ウラギンスジヒョウモン：
　Eastern Silverstripe
ウラギンスジヒョウモン：Pallas's Fritillary
ウラギンヒョウモン：High Brown Fritillary
ウラクロシジミ：Silver Hairstreak
ウラゴマダラシジミ：Blue Hairstreak
ウラジャノメ：Woodland Brown
ウラジロミドリシジミ：
　Saphirinus Green Hairstreak
ウラナミアカシジミ：Zebra Hairstreak
ウラナミシジミ：Long-tailed Blue
ウラナミシジミ：Pea Blue
ウラナミジャノメ：Japanese Rings
ウラナミシロチョウ：Mottled Emigrant
ウラベニヒョウモン：Common Leopard
ウラミスジシジミ：Alphabetical Hairstreak

エ

エゾシロチョウ：Black-veined White
エゾスジグロシロチョウ：
　Green-veined White
エゾヒメシロチョウ：Wood White
エゾミドリシジミ：Jezo Green Hairstreak
エルタテハ：Compton Tortoiseshell
エルタテハ：False Comma
エルタテハ：Large Comma

オ

オオイチモンジ：Poplar Admiral
オオウラギンスジヒョウモン：
　Great Eastern Silverstripe
オオウラギンヒョウモン：Large High Brown
オオゴマシジミ：Eastern Large Blue
オオゴマダラ：Large Tree Nymph
オオゴマダラ：Mangrove Tree Nymph
オオゴマダラ：Ocean Tree-Nymph
オオゴマダラ：Paper-butterfly
オオゴマダラ：Rice Paper Butterfly
オオゴマダラ：Tree Nymph Butterfly
オオシロモンセセリ：Grass Demon
オオチャバネセセリ：Flower Swift
オオヒカゲ：Large Brown
オオミスジ：Large Sailer
オオミドリシジミ：Oriental Hairstreak
オオムラサキ：Great Purple
オオムラサキ：Great Purple Emperor
オオムラサキ：Japanese Emperor
オオモンシロチョウ：Large Cabbage White
オオモンシロチョウ：Large White
オオルリシジミ：Large Shijimi Blue

オガサワラシジミ：Ogasawara Hedge Blue
オガサワラセセリ：Ogasawara Swift
オキナワカラスアゲハ：Okinawa Peacock
オキナワビロウドセセリ：
    Common Banded Awl
オジロシジミ：Gram Blue
オナガアゲハ：Long Tail Spangle
オナガシジミ：Walnut Hairstreak
オナシアゲハ：Citrus Swallowtail
オナシアゲハ：Lemon Butterfly
オナシアゲハ：Lime Butterfly
オナシアゲハ：Lime Swallowtail

カ

カクモンシジミ：Plumbago Blue
カクモンシジミ：Zebra Blue
カシワアカシジミ(キタアカシジミ)：
    Northern Orange Hairstreak
カバイロシジミ：Lycormas Blue
カバタテハ：Angled Castor
カバマダラ：Plain Tiger
カラスアゲハ：Bianor Peacock
カラスアゲハ：Chinese Peacock
カラスシジミ：White-letter Hairstreak
カラフトセセリ：Essex Skipper
カラフトセセリ：European Skipper
カラフトタカネキマダラセセリ：
    Northern Checkered Skipper
カラフトヒョウモン：Iphigenia Fritillary
カラフトルリシジミ：Cranberry Blue
ガランピマダラ：Common Crow Butterfly
ガランピマダラ：Common Indian Crow
カワカミシロチョウ：White Albatross

キ

キアゲハ：Common Yellow Swallowtail
キアゲハ：Old World Swallowtail
キイロウスバアゲハ：
    Eversmann's Parnassius
キイロウスバアゲハ：Yellow Apollo
キシタアゲハ：Golden Birdwing
キシタウスキシロチョウ：Orange Emigrant
キタキチョウ：Mandarin Grass Yellow
キタテハ：Asian Comma
キタテハ：Chinese Comma
キチョウ(ミナミキチョウ)：
    Common Grass Yellow
キバネセセリ：Yellow Awl
ギフチョウ：Japanese Luehdorfia
キベリタテハ：Camberwell Beauty
キベリタテハ：Mourning Cloak
キベリタテハ：White-bordered
キマダラセセリ：Japanese Dart
キマダラモドキ：Pseudo-Labyrinth
キマダラルリツバメ：Japanese Silverlines
キリシマミドリシジミ：
    Wonderful Green Hairstreak
ギンイチモンジセセリ：
    Silver-lined Skipper
ギンボシヒョウモン：Dark Green Fritillary

ク

クジャクチョウ：Peacock
クモガタヒョウモン：
    Anadyomene Fritillary
クモマツマキチョウ：Orange Tip
クモマベニヒカゲ：Arran Brown
クロアゲハ：Spangle
クロコノマチョウ：Dark Evening Brown
クロシジミ：Gray-pointed Pierrot
クロセセリ：Restricted Demon
クロツバメシジミ：Black Cupid
クロテンシロチョウ：Psyche
クロヒカゲ：Diana Treebrown
クロヒカゲモドキ：Marginalis Treebrown
クロボシセセリ：Indian Palm Bob
クロボシセセリ：Oriental Palm Bob
クロボシセセリ：Palm Bob
クロマダラソテツシジミ：
    Cycad Blue Butterfly
クロマダラソテツシジミ：Plains Cupid
クロミドリシジミ：Copper Hairstreak

付録：日本産蝶類名称の英和／和英編

コ

ゴイシシジミ：Forest Pierrot
ゴイシツバメシジミ：Moore's Cupid
コウトウシロシタセセリ：Trebellius Flat
コキマダラセセリ：Large Skipper
コジャノメ：Lilacine Bushbrown
コチャバネセセリ：Japanese Swift
コツバメ：Tailless Hairstreak
コノハチョウ：Indian Leaf Butterfly
コノハチョウ：Orange Oak Leaf
コノハチョウ：Orange Oakleaf
コヒオドシ：Mountain Tortoiseshell
コヒオドシ：Small Tortoiseshell
コヒョウモン：Ino Fritillary
コヒョウモン：Lesser Marbled Fritillary
コヒョウモンモドキ：Ambigua Fritillary
コヒョウモンモドキ：Heath Fritillary
ゴマシジミ：Scarce Large Blue
ゴマシジミ：Teleius Large Blue
ゴマダラチョウ：Japanese Circe
コミスジ：Common Sailer
コムラサキ：European Purple Emperor
コムラサキ：Freyer's Purple Emperor
コモンタイマイ：Tailed Jay

サ

サカハチチョウ：Large Map Butterfly
サツマシジミ：Albocaerulean
サツマシジミ：Blackburn's Blue
サトキマダラヒカゲ：
　　Goschkevitshi's Labyrinth

シ

シータテハ：Comma
ジャコウアゲハ：Chinese Windmill
ジャノメチョウ：Dryad
ジャノメチョウ：Dryad Butterfly
ジョウザンシジミ：Orion Blue
ジョウザンミドリシジミ：
　　Cognatus Green Hairstreak
シルビアシジミ：Lesser Grass Blue
シロウラナミシジミ：Metallic Cerulean
シロオビアゲハ：Common Mormon
シロオビヒカゲ：Bamboo Treebrown
シロオビヒメヒカゲ：Scarce Heath
シロオビマダラ：Malayan Crow
シロミスジ：Common Sergeant
シロモンクロシジミ：Apefly

ス

スギタニルリシジミ：Sugitani's Hedge Blue
スジグロカバマダラ：Common Tiger
スジグロカバマダラ：Orange Tiger
スジグロシロチョウ：Gray-veined White
スジグロシロマダラ：Black Veined Tiger
スジグロチャバネセセリ：Leoninus Skipper
スジボソヤマキチョウ：Lesser Brimstone
スミナガシ：Constable

ソ

ソテツシジミ：Cycad Butterfly ***

タ

ダイセツタカネヒカゲ：Daisetsuzana Arctic
ダイセツタカネヒカゲ：Melissa Arctic
ダイセツタカネヒカゲ：
　　White Mountain Arctic
ダイミョウセセリ：China Flat
タイワンアオバセセリ：Brown Awl
タイワンアサギマダラ：
　　Swinhoe's Chocolate Tiger
タイワンキチョウ：
　　Three Spots Grass Yellow
タイワンキチョウ：Three-spot Grass Yellow
タイワンキマダラ：Rustic
タイワンクロボシシジミ：Malayan
タイワンシロチョウ：Chocolate Albatross
タイワンツバメシジミ：Indian Cupid
タイワンツバメシジミ：Tailed Cupid
タイワンビロウドセセリ：
　　White Banded Awl

タイワンモンシロチョウ：
　Indian Cabbage White
タカネキマダラセセリ：Arctic Skipper
タカネキマダラセセリ：Checkered Skipper
タカネキマダラセセリ：Chequered Skipper
タカネヒカゲ：Asamana Arctic
タッパンルリシジミ：Pale Hedge Blue
タテハモドキ：Peacock Pansy

チ

チャバネセセリ：Small Branded Swift
チャマダラセセリ：Maculatus Skipper
チョウセンアカシジミ：Korean Hairstreak

ツ

ツシマウラボシシジミ：Blue Quaker
ツバメシジミ：Short-tailed Blue
ツマキチョウ：Yellow Tip
ツマグロキチョウ：Angulated Grass Yellow
ツマグロヒョウモン：Indian Fritillary
ツマジロウラジャノメ：
　White-tipped Woodland Brown
ツマベニチョウ：Great Orange Tip
ツマムラサキマダラ：Striped Blue Crow

テ

テツイロビロウドセセリ：Common Awl
テングチョウ：European Beak
テングチョウ：Nettle-tree Butterfly
テングチョウ：Snout Butterfly

ト

トガリチャバネセセリ：Large Banded Swift
トラフシジミ：Japanese Flash

ナ

ナガサキアゲハ：Great Mormon
ナミエシロチョウ：
　Ceylon Lesser Albatross
ナミエシロチョウ：Common Albatross
ナミエシロチョウ：Lesser Albatross

ネ

ネッタイアカセセリ：Common Palm Dart
ネッタイアカセセリ：Pale Palmdart

ハ

バナナセセリ：Banana Skipper
ハマヤマトシジミ：African Grass Blue
ハマヤマトシジミ：Dark Grass Blue
ハヤシミドリシジミ：Hayashi Hairstreak

ヒ

ヒオドシチョウ：Large Tortoiseshell
ヒオドシチョウ：
　Yellow-legged Tortoiseshell
ヒカゲチョウ：Sicelis Treebrown
ヒサマツミドリシジミ：
　Hisamatsu Green Hairstreak
ヒメアカタテハ：Painted Lady
ヒメアサギマダラ：Glassy Tiger
ヒメイチモンジセセリ：
　Small Straight Swift
ヒメウスバアゲハ：
　White Clouded Parnassius
ヒメウラナミシジミ：Common Lineblue
ヒメウラナミジャノメ：Argus Rings
ヒメウラナミジャノメ：Common Five-ring
ヒメウラボシシジミ：Quaker
ヒメギフチョウ：Small Luehdorfia
ヒメキマダラセセリ：Ochracea Skipper
ヒメキマダラヒカゲ：Small Labyrinth
ヒメシジミ：Silver-studded Blue
ヒメジャノメ：Chinese Bushbrown
ヒメシロチョウ：Eastern Wood White
ヒメチャマダラセセリ：Grizzled Skipper
ヒメヒカゲ：False Ringlet
ヒョウモンチョウ：Daphne Fritillary
ヒョウモンチョウ：Marbled Fritillary
ヒョウモンモドキ：Scotosia Fritillary
ヒロオビミドリシジミ：
　Latifasciatus Green Hairstreak

付録：日本産蝶類名称の英和／和英編

フ

フジミドリシジミ：Fujisan Green Hairstreak
フタオチョウ：Great Nawab
フタスジチョウ：East European Sailer
フタスジチョウ：Hungarian Glider

ヘ

ベニシジミ：Small Copper
ベニヒカゲ：Japanese Argus
ベニモンアゲハ：Common Rose
ベニモンアゲハ：
　Common Rose Swallowtail
ベニモンカラスシジミ：
　Red-spotted Hairstreak
ヘリグロチャバネセセリ：Sylvaticus Skipper

ホ

ホシチャバネセセリ：
　Japanese Scrub Hopper
ホシボシキチョウ：
　Broad-bordered Grass Yellow
ホシボシキチョウ：Small Grass Yellow
ホシミスジ：Dod-dash Sailer
ホソオチョウ：Dragon Swallowtail
ホソオチョウ：
　Sericin Swallow-tail Butterfly
ホソバセセリ：Silver-spotted Skipper
ホソバヒョウモン：Thor's Fritillary
ホソバヒョウモン：Thore Fritillary
ホリイコシジミ：Tiny Grass Blue

マ

マサキウラナミジャノメ：Masaki's Rings
マルバネウラナミシジミ：
　Dingy Line Blue
マルバネウラナミシジミ：Dingy Lineblue
マルバネルリマダラ：
　Blue Branded King Crow
マルバネルリマダラ：
　Blue-branded King Crow

ミ

ミカドアゲハ：Common Jay
ミズイロオナガシジミ：
　Black-banded Hairstreak
ミスジチョウ：Long-streak Sailer
ミスジチョウ：Sailer
ミダムスルリマダラ：Blue Spotted Crow
ミダムスルリマダラ：Blue-spotted Crow
ミドリシジミ：Green Hairstreak
ミドリヒョウモン：Silver-washed Fritillary
ミナミコモンマダラ：Blue Tiger
ミヤマカラスアゲハ：Maackii Peacock
ミヤマカラスシジミ：Mera Black Hairstreak
ミヤマシジミ：Argyrognomon Blue
ミヤマシジミ：Reverdin's Blue
ミヤマシロチョウ：
　Oriental Black-veined White
ミヤマセセリ：Spring Flat
ミヤマチャバネセセリ：Janson's Swift
ミヤマモンキチョウ：
　Alpine Clouded Yellow
ミヤマモンキチョウ：
　Moorland Clouded Yellow

ム

ムモンアカシジミ：Jonasi Orange Hairstreak
ムラサキオナガウラナミシジミ：
　Forget-me-not
ムラサキシジミ：Japanese Oakblue
ムラサキツバメ：Powdered Oakblue
ムラサキテングチョウ：Purple Beak

メ

メスアカミドリシジミ：
　Smaragdinus Green Hairstreak
メスアカムラサキ：Danaid Eggfly
メスグロヒョウモン：Sagana Fritillary

モ

モンキアゲハ：Red Helen

モンキチョウ:
　Eastern Pale Clouded Yellow
モンキチョウ: Pale Clouded Yellow
モンシロチョウ: Cabbage White
モンシロチョウ: Small Cabbage White
モンシロチョウ: Small White

ヤ

ヤエヤマイチモンジ: Staff Sergeant
ヤエヤマウラナミジャノメ:
　Yayeyamana Rings
ヤエヤマカラスアゲハ:
　Yayeyamana Peacock ***
ヤエヤマムラサキ: Crow Eggfly
ヤエヤマムラサキ: Malayan Eggfly
ヤクシマルリシジミ: Common Hedge Blue
ヤマキチョウ: Brimstone
ヤマキチョウ: Sulphur
ヤマキマダラヒカゲ: Japanese Labyrinth
ヤマトシジミ: Pale Grass Blue
ヤマトスジグロシロチョウ:
　Green-veined White

ユ

ユウレイセセリ: Formosan Swift
ユウレイセセリ: Rice Swift

リ

リュウキュウアサギマダラ:
　Ceylon Blue Glassy Tiger
リュウキュウウラナミジャノメ:
　Riukiuana Rings
リュウキュウウラボシシジミ:
　Forest Quaker
リュウキュウヒメジャノメ:
　Riukiuana Chinese Bushbrown
リュウキュウミスジ: Hylas Common Sailer
リュウキュウムラサキ: Common Eggfly
リュウキュウムラサキ: Great Eggfly
リュウキュウムラサキ: Jacintha Eggfly
リュウキュウムラサキ: Varied Eggfly
リンゴシジミ: Black Hairstreak

ル

ルーミスシジミ: Tailless Bushblue
ルリウラナミシジミ: Dark Cerulean
ルリシジミ: Hill Hedge Blue
ルリシジミ: Holly Blue
ルリシジミ: Sivery Hedge Blue
ルリタテハ: Blue Admiral
ルリマダラ: Double-branded Crow
ルリマダラ: Two-brand Crow

おわりに

　インターネット時代に冊子体である本書は、調べる上で、学ぶ上で有用でしたでしょうか。その際に、少しでもお役に立てれば著者として望外の喜びです。

　この『蝶類生物学英和辞典』は、門外漢が作成した辞典で、インターネット時代だからこそ編纂ができた辞典です。

　自然科学を深く学ぶには専門用語に関する英語の知識が必須です。自然科学 → 生物学 → 昆虫学 → 蝶類学と専門分野が特化すればするほど、より特化した専門英語辞典が必要となります。

　生物学に関する専門用語辞典（英語辞典／英和辞典）は内外に多々あり、それらより専門範囲が狭い昆虫学に関する専門用語辞典も内外には多少あります。しかし、それらは、さらに狭い専門領域である「蝶類」に関連する英語版の文献を読んだり、書いたりする際には、余り役に立たないのが現状です。

　それならば、蝶類関連の英語版の論文や記事、図鑑、専門書を読んだり書いたりする際に、本当に役立つ実用的な真の専門英和辞典を自身で編纂しようと思い立ち、作成を試みたのが本書です。

　そこで、本書では「蝶類の生物学」の学問範囲である「分類、生理、生態、形態、遺伝、発生、病理、行動、保全、採集」の各領域をできる限り網羅し、さらに、現代の研究潮流である最新の分子生物学の専門英用語をもかなり網羅するように努めました。

　私は 10 年前に定年退職したコンピュータ技術者で、蝶採集の長年の愛好家ですが、生物学や昆虫学、蝶類学に関しては素人です。しかし、現代はインターネット時代であり、必要な情報・知識は、大学や研究機関に所属して

いなくても、無料で迅速に入手できる環境になっています。特に、自然科学分野はオープンサイエンスの時代となり、自然科学系学会誌や一部の市販科学雑誌のほぼ最新の論文や記事、ブログ記事などの電子版を無料で入手できる環境になっています。このようなインターネット環境の中でコンピュータを駆使して、8,800語句の蝶類を主体とする生物学の英和辞典を作成しました。

特に、スペルチェックに関連して言えば、英語関連の辞典の要諦は、英単語のスペルの正確さですので、スペル誤りの撲滅に注力しました。目視チェックとコンピュータ及びインターネットを活用したスペルチェックを徹底的に行いました。本書『蝶類生物学英和辞典』は世界で初めて刊行される辞典ですので、収載される英単語・用語の大半は、「蝶類」分野特有の最新の専門語句です。そのため、複数の著名で大規模なオンライン版スペルチェッカー（例えば、生命科学用のWebSpell、汎用のオンライン版MicroSPELLなど）でチェックすると、当初、収載の全英単語の約15％にスペル誤りの可能性があると指摘されました。これら15％の英単語は徹底的に調べ、スペル誤りが確信された英単語はスペルを訂正しました。

一双である付録『日本産蝶類名称の英和／和英編』に関して言えば、インターネット上で"学名（属名 種名）" & "common name" を、例えば、"Sasakia charonda common name" をキーワードとして検索しさえすれば、簡単に短期間で作成できると思っていましたが、いざ作成に取り掛かると、すぐにそう簡単ではないことに気付きました。

唯一つの有効な「学名」では生じませんが、一般的呼称である「英名」には、「同物異名」と「異物同名」の問題が発生しました。複数個の「同物異名」に関しては、その蝶の学名の「亜種」水準まで厳格に吟味し、かなりの個数の「異名」を候補対象から削除しました。「異物同名」に関しては、異物の「別名」をインターネット上は勿論のこと、それ以外の資料類でも探し、この問題を極力回避しました。

今日のインターネットは「情報の宝庫」です。しかし、生来的な問題点の

おわりに

一つとして言われる「インターネットはゴミだらけ」の証左でもある過去の記載誤りが、インターネット上で消えずにそのまま蓄積・通用しているという現状があります。その例が、「オオイチモンジ」の「Popular Admiral（正しくは、Poplar Admiral）」です。

　最終的に英名を選定・決定するには、各蝶に関する亜種水準までの学名と同時に、その蝶の形態面・生態面に関する深い専門知識が必要であると、実感しました。

　この日本産蝶類の英名に関して、読者諸兄姉からご意見・指摘を頂ければ幸いです。

　ニューサイエンス社編集部の角谷裕通氏には、出版不況と言われるなか、インターネット時代に冊子体の辞典類の出版は採算が合わず、同時にインターネットの利便性や圧倒的な検索能力には勝てないと一般には思い込まれている趨勢下で、このニッチな辞典の出版を引き受けて頂いたことに敬意を表します。

　最後になりましたが、監修者である日本大学生物資源科学部の岩野秀俊教授には、貴重な夏休みの2ヶ月間（平成26年の夏）を充て、蝶類学・昆虫学の専門的立場から精力的かつ克明に校閲して頂きました。これにより著者の専門知識の欠如が補われ、同時に「インターネットはゴミだらけ」と揶揄される本書の情報ソースの弱点の克服に大きな助力を得たことを記し、ここに謝意を表します。

平成27年2月

　　　　　　　　　　　古希を記念して　　鍛治勝三　記す

《監修者略歴》

岩野秀俊（いわの・ひでとし）

博士（農学）（九州大学農学部）

1951年　東京都に生れる。
1975年　日本大学農獣医学部農学科卒業。
1978年　日本大学農獣医学部助手。
1993年　ミネソタ大学農学部昆虫学科客員研究員。
2006年　日本大学生物資源科学部教授。

専門分野　応用昆虫学・昆虫病理学・害虫管理学

学会活動　日本鱗翅学会理事、日本応用動物昆虫学会評議員、日本蚕糸学会代議員、相模の蝶を語る会代表、Member of Lepidopterists' Society（USA）

著　書　バイオロジカルコントロール（朝倉書店）、天敵微生物の研究手法（日本植物防疫協会）、微生物の事典（朝倉書店）、微生物の世界（筑波出版会）、微生物の資材化：研究の最前線（ソフトサイエンス社）、日本の昆虫の衰亡と保護（北隆館）、日本産蝶類の衰亡と保護（日本鱗翅学会）、神奈川県昆虫誌（神奈川昆虫談話会）、相模原市史自然編（相模原市総務局）

《著者略歴》

鍛治勝三（かじ・かつぞう）

1944年　東京都に生れる。
1967年　法政大学工学部電気工学科卒業。
　同年　財団法人日本情報処理開発センター（JIPDEC）に入社。
2004年　独立行政法人情報処理推進機構（IPA/JITEC）に転籍。
　同年　定年退職。

現　在　退職後は、フリーランスとして、自宅で電子メールによるやり取りのみで国際規格（ISO）の作業暫定版の英文和訳の仕事に従事。

趣　味　蝶の採集で、定年退職前は東京都や山梨県が採集フィールドでしたが、退職後に初めて行った（神奈川県の）湘南方面でモンキアゲハやナガサキアゲハなどの黒色系アゲハがたくさん飛来していたことに驚きました。また、ちょうどその頃、外来移入種であるアカボシゴマダラがたくさん飛んで来るヒルトッピングの場所を探しあて、今も駆除（採集）しています。

グリーンブックス
蝶類生物学英和辞典

平成 27 年 3 月 20 日　初版発行
〈図版の転載を禁ず〉

監　修	岩　野　秀　俊
著　者	鍛　治　勝　三
発行者	福　田　久　子
発行所	株式会社 ニューサイエンス社

〒108-0074　東京都港区高輪3-8-14
電話03(5449)4698　振替00160-9-21977
http://www.hokuryukan-ns.co.jp/
e-mail：hk-ns2@hokuryukan-ns.co.jp

印刷所　　株式会社 東邦

© 2015 New Science Co.
ISBN978-4-8216-0140-0 C0345

当社は、その理由の如何に係わらず、本書掲載の記事（図版・写真等を含む）について、当社の許諾なしにコピー機による複写、他の印刷物への転載等、複写・転載に係わる一切の行為、並びに翻訳、デジタルデータ化等を行うことを禁じます。無断でこれらの行為を行いますと損害賠償の対象となります。

また、本書のコピー、スキャン、デジタル化等の無断複製は著作権法上での例外を除き禁じられています。本書を代行業者等の第三者に依頼してスキャンやデジタル化することは、たとえ個人や家庭内での利用であっても一切認められておりません。

連絡先：ニューサイエンス社 著作・出版権管理室
Tel. 03(5449)7064

JCOPY 〈(社)出版者著作権管理機構 委託出版物〉
本書の無断複写は著作権法上での例外を除き禁じられています。複写される場合は、そのつど事前に、(社)出版者著作権管理機構（電話：03-3513-6969, FAX:03-3513-6979, e-mail: info@jcopy.or.jp）の許諾を得てください。